The Barbary Macaque:
Biology, Management and Conservation

Cover photo: J.K. Hodges

Artist: Vicky White

The Barbary Macaque: Biology, Management and Conservation

Edited by

J.K. Hodges

J. Cortes

NOTTINGHAM
University Press

First published by Nottingham University Press
This reissued original edition published 2023 by 5m Books Ltd www.5mbooks.com

British Library Cataloguing in Publication Data
The Barbary Macaque: Biology, Management and Conservation
I. Hodges, J.K. II. Cortes, J.

ISBN 9781789183115

Disclaimer

Typeset by Nottingham University Press, Nottingham

EU GPSR Authorised Representative
LOGOS EUROPE, 9 rue Nicolas Poussin, 17000, LA ROCHELLE, France
E-mail: Contact@logoseurope.eu

Preface

The last publication dedicated to the Barbary macaque was the volume edited by John Fa and published more than 20 years ago, in 1984. It is no coincidence that the editor of that much referenced work should have been from Gibraltar, nor indeed that one of the editors of this one should also hail from the Rock.

While Gibraltar cannot, and does not claim to be the original home of the Barbary macaque, it is nevertheless true that the species' rise to fame, certainly outside the realm of academia, is directly attributable to the touristic exploitation of this species' outpost in southern Iberia.

The accessibility of this population to study, as far back as the 1960s and 1970s, allowed Burton and Sawchuk to carry out their demographical work, which was continued and expanded upon later by Fa. In the early 1990s, the Gibraltar Ornithological & Natural History Society (GONHS) was diversifying from its original interest in ornithology and developing a growing interest in research-driven management and conservation practices. At about this time, it was Robert Martin, then of the Anthropological Institute of Zurich, who made contact with GONHS, establishing the first collaborative programme within what was to later become the Gibraltar Barbary Macaque Project, which now not only includes the original genetic studies, but which also encompasses physiological, behavioural and ecological studies involving many other research institutions. It was this expanding international network, with its improved contacts to macaque researchers and managers, including those of the host countries of Morocco and Algeria, that led ultimately to the Calpe Conference on the Barbary macaque held in Gibraltar at the end of 2003. And, as often occurs, it was the conference which identified the need and provided the inspiration for this second publication dedicated to the Barbary macaque.

The book, like the Conference, covers a wide range of subject matter. Our intention has been to look at the species from various perspectives and to do so in a way that would be useful to both primatologists and managers alike, and with appeal to both a specialist and non-specialist readership. In this way, the work provides information of both historical and academic interest as well as of practical value. Although it is first and foremost about the Barbary macaque, there is also a strong comparative element in which the species´ relationship to other primates is repeatedly examined.

By way of introduction, the first part of the book provides a historical perspective, covering origins, dispersal patterns and phylogenetic relationships within the macaque genus, as well as highlighting the special position the Barbary macaque has held in its relationship to Man.

The principal aim of the second part of the book is to present a synthesis of current knowledge of the biology of the Barbary macaque and in doing so to recognise the contribution it has made to our understanding of primate socio-biology in general. Looking at the species from a comparative perspective within the macaque genus as a whole, chapters also highlight key differences and similarities between the Barbary and its Asian cousins in order to examine what these could be telling us about evolutionary processes shaping macaque (and primate) socio-biology on a broader scale.

Unfortunately, the Barbary macaque, like an increasing number of other primate species, faces a uncertain future. Officially classified as "vulnerable", the status of the species in the wild is in fact progressively deteriorating, its distribution now limited to a few isolated populations in the mountainous regions of Morocco and Algeria. The second objective of the book (part 3) therefore, is to briefly review information relevant to the management and conservation of the species. Looking at both captive (including semi-captive or range-limited) and wild populations, chapters deal with genetic management, husbandry and veterinary care as well as status updates, demographic trends and the perennial problem of human-animal interactions. The Gibraltar macaques are, not surprisingly, looked at closely, as are wild macaques in Morocco, whose conservation status currently gives cause for particular concern. It is hoped that this section of the book will be useful, not only in providing practical information useful towards achieving more effective species management, but also, by (re-) focussing attention on the problems facing the Barbary macaque, in highlighting the urgency with which action is required if its long-term survival is to be ensured.

While there are still gaps in our knowledge, we believe this book provides a useful up to date synthesis of what we do know about the Barbary macaque. We hope that it will also help to cement a working relationship between its many contributors and between them and others in this field, and we look forward to continuing developments leading to, as was proposed in 2003, a further Meeting on the species several years hence.

Acknowledgements

The editors would like to thank the Government of Gibraltar for patronage of the Calpe 2003 Conference, *The Barbary Macaque: Comparative and Evolutionary Perspectives*, which led to the publication of this book, and in particular to the then Minister for the Environment The Hon Keith Azopardi, for his inaugural address and his initiative in setting up the Calpe conference series.

Funding for the Conference and for the publication of this book was made available by the Government of Gibraltar and the European Union through GIBMANATUR, a project under the Gibraltar-Morocco Interreg IIIA Programme. We would like to thank the Gibraltar Government's EU funding Unit for their assistance, in particular Francis Sheriff, Charles Collinson, Jessica Harrison, James Savigon, Douglas Ryan and Jason Davies.
We would also like to acknowledge the German Primate Centre for providing funds to support a travel bursary scheme for student attendees of the Conference.

Sincere thanks are due to the Gibraltar Ornithological & Natural History Society, the German Primate Centre, and the Institute Scientifique of the University Mohammed V Rabat-Agdal for their support in organising the Conference.

The editors also wish to acknowledge the contribution of all the authors, as well as their research teams and assistants. We would also like to thank Sarah Keeling at Nottingham University Press for her assistance with the preparation of this volume and Ellen Wiese for her invaluable secretarial support throughout the editing process.

CONTENTS

Part 3: Management and Conservation

From Calpe to Catanduanes: Babewynes, apes, marmesettes and othere dyverse bestes

C Groves
School of Archaeology & Anthropology, Australian National University, Canberra, Australia

Macaca is the most widespread genus of primates after *Homo*. Comprising at least 25 species, it lives in Europe (just) and North Africa, and after a gap it occurs throughout South and Southeast Asia and North into Japan. Gibraltar and the Philippines do not correspond exactly to the western- and eastern-most points of macaque distribution, but 'Calpe to Catanduanes' fairly well represents the extraordinarily wide spread of the genus.

This brief survey of the genus *Macaca* will be broadly divided into three sections:

Barbary macaques in history: how macaques in general, and Barbary macaques in particular, have been regarded through history in the western world.

How Barbary macaques got their names: nomenclature of macaques, in particular the Barbary macaque.

The taxonomy and phylogeny of macaques: taxonomy more or less depends on phylogeny, so these two topics will be treated together. I will briefly discuss both how the genus *Macaca* is related to other old world monkeys, and how the species of the genus are related to each other.

Barbary macaques in history

The Ancient Egyptians originally knew and depicted three species of cercopithecoid monkey: the Hamadryas baboon (*Papio hamadryas*), Vervet monkey (*Chlorocebus aethiops*), and probably the Gelada (*Theropithecus gelada*).

The Hamadryas baboon represented the God Thoth, the god of writing, although in art Thoth was more usually depicted as an ibis and the baboon as assistant. A baboon is seen guarding the first gate of the underworld, or forms the stopper of the canopic jar containing the lungs of a mummified dignitary. Because hamadryas baboons face the rising sun, bark and screech to it, and stretch out their arms to welcome it, they were sacred to the sun god, and a hamadryas was sometimes shown bearing the solar disk on his head.

Baboons were much admired for their intelligence. They were kept as pets, and they were mummified. The necropolis of Hermopolis contains thousands of mummified ibises and baboons. It is quite likely that in Pharaonic times hamadryas lived in Egypt itself, so we need not ask whence they got these large numbers.

Vervet monkeys were less sacred, but also frequently kept as pets. They were evidently traded overseas, and there are frescoes of them from Minoan sites on Crete and on Thera.

The gelada was less well-known in ancient Egypt than the hamadryas or the vervet, but may have had an even greater impact: it was perhaps the Sphinx. In a classic paper, Jolly & Ucko (1968) showed that, on the rare occasion when a gelada was caught and brought to Egypt, it was described as a 'sphinx'. This of course does not necessarily mean that the great Sphinx of Giza was based on a gelada, but the similarities spring out at us when we read Jolly and Ucko's paper (1968), and unless concepts like this are invented out of whole cloth, the gelada would have been a better model for the Sphinx than anything else I can think of.

The Maghreb was just along the coast to the west, but it seems that in the main the Egyptians were interested in going south, down the Nile, because Barbary macaques took a very long while to enter the Egyptians' sphere. I know of no depictions of them in Pharaonic art, and not until the Ptolemaic era is there any indication that the Egyptians knew about them.

In 1972, I was asked to examine the skulls of 16 mummified 'baboons' from the Ptolemaic-era catacombs at Saqqara, dating from after 300 B.C. I was surprised to find no hamadryas; instead I identified twelve anubis baboons, two vervets - and two Barbary macaques.

The study was for an intended book celebrating the work of the British Egyptologist Walter Bryan Emery (1903-1971), but the book was never published, because of the unexpected death of the editor. In the 1990s there was a proposal to revive studies of Saqqara and its mummies, and so it was that in 1996 the noted British primatologist Doug Brandon-Jones visited Saqqara on an expedition mounted by the Egypt Exploration Society of the University of Amsterdam.

The results have been briefly published (Goudsmit & Brandon-Jones, 1999). Brandon-Jones studied 169 skulls, and confirmed my own findings: 149 were baboons, but *Papio anubis*, **not** *P.hamadryas*; there were however no vervets — the remainder were all Barbary macaques, *Macaca sylvanus*.

Presumably this means that, under its new Greek Pharaohs, Egypt had begun to be more outward looking than before.

The world was changing fast. The Carthaginians traded widely, and presumably it was from this trade that a Barbary macaque ended up at Navan Fort, Armagh, about 150 B.C. (Lynn, 1997). Rome destroyed Carthage in 146 B.C. and colonised the Maghreb. Barbary macaques attracted their attention, and became widely kept as pets throughout the Roman Empire and allied regions. Their remains are known from several sites in Western Europe from the Roman era:

- Pompeii, Italy, 79 A.D. (Bailey *et al.*, 1999)
- Dunstable, Bedfordshire, England, late 2nd. century A.D. (Lynn, 1997).
- Catterick Fort, Yorkshire, England, "probably Romano-British" (Lynn, 1997).

Barbary macaques were becoming known throughout the Mediterranean world in another respect, too. Galen of Pergamum (Claudius Galenus), who lived between about A.D. 129 and 203, was a Greek philosopher who became physician to the Roman Emperor Marcus Aurelius. He was of an inquiring disposition, and rather unusual among philosophers of the time in his practical bent; it did not suffice to him merely to think about the world, but he set about trying to discover it practically. As he was making his living as a physician, it was to anatomy that he turned. Material, however, was a problem. He was able to dissect a few corpses of robbers and of gladiators killed in the arena, but mainly he had to use Barbary macaques as proxy for humans. He wrote:

> Choose those apes likest man, with short jaws and small canines. You will find other parts also resembling man's, for they can walk and run on two feet. Those, on the other hand, like the dog-faced baboons, with long snouts and large canines, far from walking or running on their hind-legs, can hardly stand upright.

Galen, *On Anatomical Procedure* (Book I, Chapter II)

Unfortunately, his inquiring spirit did not persist among his successors, and right up until the 16th century in Europe his works were consulted as the last word on human anatomy, and few would dare to question them. That they were based mainly on dissection of Barbary macaques was either lost on mediaeval philosophers and physicians, or was regarded as of no account. This is the attitude that held up medical advances for over 1000 years, and things did not change until Andreas Vesalius in the 1530s dared to challenge

Galen's Holy Writ, and performed his own dissections of human cadavers. But that is another story.

In Mediaeval Europe, Barbary macaques were generally referred to as 'apes'. The difference between 'apes' and 'monkeys' was vague at best, but as long tailed monkeys increasingly began to be imported into Europe, either through trade with North Africa and from Portuguese voyages to West Africa and Brazil, the term 'ape' increasingly began to be restricted to the tailless Barbary macaque. (The derivation of 'ape' is obscure; it first occurs as early as about 700 A.D. The term 'monkey', which does not appear in English until about 1600, may be a shortening of 'mannekin' meaning little man). Galen had used Barbary macaques as proxy humans; in the Middle Ages in Europe they continued in this role, but in quite a different way. They were often symbols of the lecherous side of humanity, or of the devil. Sometimes they were regarded more as symbols of the exotic: Sir John Mandeville, in 1366, describing what purported to be his own voyage to the mythical Christian Empire of Prester John (though, beyond the Holy Land, which he probably did actually visit, it is almost certainly entirely fictional), described how an island was inhabited by Babewynes, apes, marmesettes and othere dyverse bestes (Baboons, apes, marmosets and other diverse beasts). This seems to have been the first use in English of the words 'baboon' and 'marmoset' – though whether he knew what baboons were is dubious, and he certainly did not know what marmosets were (the word seems to mean a humanoid creature like a gargoyle, 'a small being made of marble').

The role of Barbary macaques as creatures that were theologically special, because they were nearly but not quite human, persisted throughout the Middle Ages and even beyond. In some schools of thought they were the Devil's attempt to mock God's creation; as Man was created in God's image, so the Ape was created in the Devil's. As late as 1626 (though written more than half a century before), we read in William Roper's *Life of St.Thomas More*:

> Whosoever will mark the devil and his temptations, shall find him therein much like to an ape.

But, even by that time, more realistic views of the animal world were seeping through.

Barbary macaques (and their relatives) get their names

The ancient Greeks knew two sorts of monkeys, apart from baboons and occasional sphinxes. Aristotle and Strabo wrote of *kebos*, which in Latin spelling becomes *cebus*; in the hands of Diodorus Siculus this became *kephos*, which Latinises as *cephus*. Although it is likely that the word derives from the Egyptian GIF (for the vervet), its derivation was rationalised by Agatharchides as from *kepos*, a garden, on the grounds that the vervet is brightly coloured. Another form of the word was *keipon*, which could just possibly be the source of 'gibbon'.

The second kind of monkey in Greek writings was *pithekos*, which Latinises as *pithecus*. And this one, whenever the species can be identified, seems to be the Barbary macaque.

For the Romans, Barbary macaque was *simia*, 'the snubnosed one', from *simus* = snub-nosed. When the word for an animal has clear meaning in the same language, it implies the animal was unfamiliar and a descriptive name had to be invented for it. As we have seen, Barbary macaques probably did not enter Roman consciousness until around the time of the conquest of Carthage.

Mediaeval England had, as we have seen, the words Ape, Monkey and Marmoset. Mediaeval France had variants of these terms, and went one better: the Barbary macaque seemed to remind the French irresistibly of a wrinkled, wise old man, and so came always to be called *Magot*, from Greek/ Latin *magus* = a sage (originally an astrologer in the Persian empire), a name that persists to this day.

The multivolume *Histoire Naturelle Générale et Particulière avec la Description du Cabinet du Roi* by Georges Louis Leclerc, Comte de Buffon (1707-1788) marks a transition between the mediaeval view of 'apes', as mockeries of humanity, and a modern scientific view. Buffon could not help but remark on monkeys' ugliness, lasciviousness, and so on, but at the same time he described many new species (though not giving scientific names – he did not believe in that sort of thing), and it is generally possible, either from his descriptions or from the woodcuts that accompanied the work, to tell what species he was describing. Vol.14, published in 1766, contains descriptions of primates, including *Le Magot*, very clearly the Barbary macaque – he even, correctly, says that the tail has no vertebrae, though this little snippet may well have been due to his collaborator Daubenton, who appended brief but accurate anatomical descriptions to most of Buffon's more impressionistic ones.

Another species described by Buffon was *Le Macaque* – the first time the word had appeared in print, at least in this form. 'Macaque', he said, was 'nom de cet animal dans son pays natal, à Congo'! - as source of this statement, he cited Georges Marcgrave (1610-1643), who in his posthumous *Historia Naturalis Brasiliae* (1648) described *Cercopithecus angolensis major, Congensibus Macaquo*. Buffon accompanied his description of *Le Macaque* with a very passable engraving of what is clearly *Macaca fascicularis*, but what was Marcgrave's 'Congolese macaquo' from which it was derived?

In Linggala, the lingua franca of the Congo basin, the usual word for any monkey is *kako*. Linggala is essentially a trade creole formed out of a number of indigenous Bantu languages, and it is characteristic of the Bantu language group that the words are inflected by adding not a suffix, as in Indo-European languages, but a prefix. In Central African Bantu languages the prefix for the plural is *ma-*. If *kako* means one monkey, therefore *makako* means monkeys.

It is usual that indigenous languages have different words for different species, whereas trade languages and other creoles generalise, and have a single word that covers a range of similar species. So while in Linggala *kako* is used for any old monkey, one might expect its original form, in one or more indigenous languages, to denote one specific kind of monkey. And indeed Malbrant and Maclatchy (1949) list the word *kaku* in the following languages of the former French Equatorial Africa (in other words, Gabon, Congo Republic and Central African Republic): Munukutuba, Eshira, Bapunu and Bakota. And it means - mangabey! So a macaque is actually a mangabey…

The founder of taxonomy, including the binomial system, was Carl Linnaeus (1707-1778). The starting-point of zoological nomenclature is the 10th edition of his *Systema Naturae per Regna Tria Naturae, secundum Classes, Ordines, Genera, Species, cum characteribus, differentiis, synonymis, locis,* and in this he erected a genus, *Simia*, for all nonhuman primates except lemurs, and created three divisions in it according to the length of the tail. One of his two species was *Simia sylvanus*, based on descriptions from earlier sources, the first of which (on which his description is mainly based) is clearly the Barbary macaque. In the 12th edition of this work, the last before his death, Linnaeus described another species, *Simia inuus*, which he said resembles *Simia sylvanus* but '*rostro productiore, colore pallidiore*'. The sources which Linnaeus cited for the second species likewise described Barbary macaque. So 12 years after the Barbary macaque received its first scientific name, it received a second.

Sylvanus and Inuus were two Roman forest gods, both of them being in effect cleaned-up versions of the lecherous Greek god Pan. Even in the hands of the father of systematics, therefore, Barbary macaques were still keeping some of the aura which surrounded them in mediaeval days.

In 1799, Citizen Lacépède produced a new classification of mammals, naming several new genera. Among other things, he broke up the old Linnean genus *Simia* into several different genera; one of these was *Macaca*, with *Simia inuus* Linnaeus, 1766 as type species. So the Barbary macaque now had its genus and its species.

The taxonomy and phylogeny of macaques

How to define Macaca?

I asked this question previously (Groves, 2001), and admitted that while the

DNA evidence is decisively in favour of the monophyly of *Macaca*, I could find no convincing synapomorphies (i.e. shared derived conditions). Compared to other old world monkeys, macaques really are a varied lot: macaques are brown (except for the ones which are black), short-tailed (except for the ones with long tails), and short-faced (except for the ones with long faces); the females develop periodic sexual swellings (except for the ones which don't); and they lack molar flare (except for the ones which possess it).

Most recently, however, a series of papers by Koppe and colleagues (see, for example, Koppe & Ohkawa, 1999) has made the case that the macaques do share one noticeable synapomorphy. Old world monkeys are unique in lacking maxillary sinuses; this is most parsimoniously regarded as a loss. *Macaca*, however, possesses them; it is the only genus of Cercopithecoidea to do so. The redevelopment of the maxillary sinuses therefore would be a synapomorphy.

The discovery of maxillary sinuses in *Macaca* by Koppe and his coworkers, however, was confirmed only for Asian species. As the Asian species form a clade with respect to *Macaca sylvanus*, the possibility remains that the synapomorphy defines just the Asian clade, and not the genus as a whole. The maxilla of *Macaca sylvanus* certainly appears as inflated as that of any other macaque; to check this, I examined skulls in the collection of the Natural History Museum, London (formerly the British Museum (Natural History)), and discovered a skull conveniently broken at the back of the maxilla, revealing a large sinus. Consequently, the redevelopment of maxillary sinuses constitutes a convincing anatomical synapomorphy for the entire genus *Macaca* – the only one, so far, that seems confirmed.

The species of Macaca

For Fooden (1976 and elsewhere), the living species of macaques fall into four species groups. For their distinction, penis morphology is the key tool. His four species-groups are:

The *Macaca silenus-sylvanus* group; the *Macaca fascicularis* group; the *Macaca sinica* group; and, in a group by itself, *Macaca arctoides* (Figure 1).

It is evident from molecular work that the first of these groups is paraphyletic; *Macaca sylvanus* shares with the other species of this group only symplesiomorph conditions, notably of course the form of the penis. The other groups seem to be monophyletic, as does the *Macaca silenus* group (the Asian component of Fooden's *silenus-sylvanus* group).

In listing the species that belong under each of the five species groups, I employ the Phylogenetic Species Concept, for reasons argued previously (Groves, 2001). Under this concept, a species is -

an irreducible cluster of organisms (within which there is a parental pattern of ancestry and descent) that is diagnosably distinct from other such clusters by a unique combination of fixed heritable characters.

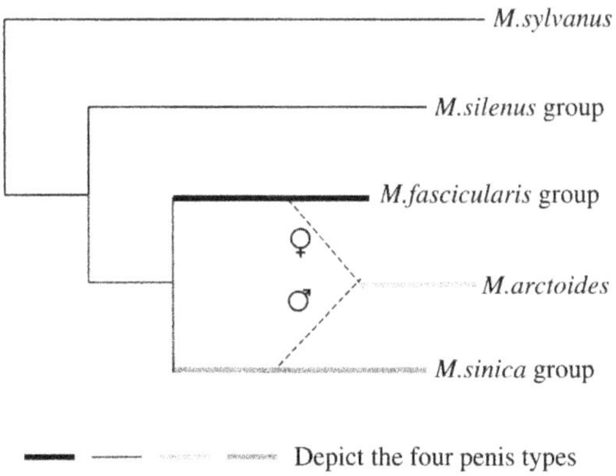

Figure 1. Macaque phylogeny (after Tosi *et al.*, 2000)

Macaca silenus group: *Macaca silenus, leonina, nemestrina, pagensis, siberu,* and the Sulawesi species, probably 8 in number (see Froehlich, this volume). The distribution is continuous throughout mainland Southeast Asia and south into Sundaland (Sumatra, Bangka and Borneo) from which it extends across deep channels into the Mentawai islands to the west and Sulawesi and offshore islands to the east. Further west on the mainland, there is a wide gap and the group occurs again in the Western Ghats of India.

According to the analysis of Morales and Melnick (1998), the *silenus* group separated from other Asian macaques shortly after the arrival of the genus in Asia about 5.5 million years ago, and the Sulawesi group separated from the others some 4.5 million years ago. Rosenblum *et al* (1997) found that *M. leonina* and what they referred to as *M. pagensis* (now known to be actually a different species, *M. siberu*) are about equally divergent from *M. nemestrina.* Because at that time it was universal to regard *leonina* as merely a subspecies of *M. nemestrina*, they used this as a reason to keep *pagensis* also as a subspecies of *M. nemestrina.* Groves (2001) not only endorsed specific status for *M. pagensis*, but argued strongly that *M. leonina* itself is a very distinct species (with quite a narrow zone of hybridisation with *M. nemestrina* on the Isthmus of Kra). The status of *Macaca pagensis* as a distinct species was likewise supported by Abegg and Thierry (2002a) on ethological grounds (in fact, like Rosenblum *et al.*, 1997, what they were studying is now known to have been *M. siberu*).

Mentawai macaques were first shown to consist of not one but two distinct species (*M. pagensis* and *M. siberu*) by Kitchener and Groves (2002); almost simultaneously, Roos *et al.* (2003) corroborated this based on molecular genetic data and further showed that the two are probably more closely related to *M. nemestrina* than to each other.

Abegg and Thierry (2002b) made a number of interesting observations about the *Macaca nemestrina* group. The peripheral species – *Macaca silenus*, the Mentawai macaques, and the Sulawesi macaques – share both morphological (mainly pelage) and ethological features to the exclusion of *M. nemestrina* and *M. leonina*. They appealed to a centrifugal mechanism (as described by Groves, 1989) to help explain this. In addition, the Pleistocene saw a series of alternate expansions and contractions of rainforest in South and Southeast Asia, and it was during the contractions that the various species of the group would have differentiated.

Morphologically, *M. nemestrina* is more highly autapomorphic than is *M. leonina*, as noticed by Abegg and Thierry (2002b). As a hypothesis, this may be related to its sympatry with the most widespread and versatile species of macaque, *M. fascicularis* (ie it has diverged more in the presence of this ecologically non-specialised competitor).

Macaca fascicularis group: *Macaca fascicularis, mulatta, cyclopis, fuscata.* The distribution of the group is continuous from the central and northern parts of South Asia northeast through China to Japan, Taiwan and Hainan, and south into most of Indonesia and the Philippines. Of the four species, *M. fascicularis* is sister to the other three. As reconstructed by Morales and Melnick (1998), the group as a whole separated from the closely related *M. sinica* group somewhat over 3.5 million years ago, and the ancestors of *M. fascicularis* diverged from the common ancestor of the other three species 2.5 million years ago.

Despite the fact that they are not themselves sister species, *M. mulatta* and *fascicularis* interbreed along a line running through Burma, Thailand, Laos and Vietnam. The hybrid zone is asymmetrical: there are more *mulatta* features on the *fascicularis* side of the zone than the converse. This situation is illustrated in Figure 2, which is based largely on Fooden (1995, 2000 and other sources) and Tosi *et al.* (2000); see also Cronin *et al.* (1980) and Eudey (1980). The hypothesis here put forward to explain this is as follows:

> *M. fascicularis* is a Sundaic species; it differentiated as the southern representative of its species group, some two and a half million years ago, well before the three northern species had begun to separate from each other. During (most of?) the Plio-Pleistocene, what is now the Isthmus of Kra was a seaway, and Sundaland was an island (Woodruff, 2003), so the two proto-species were isolated on either side of a notable

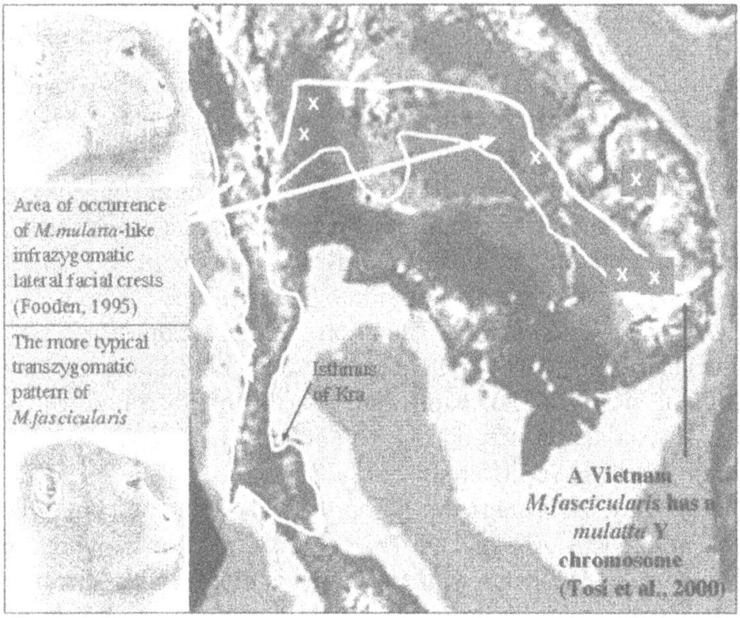

Figure 2. Distribution of the *Macaca fascicularis* group on mainland Southeast Asia, showing the characters indicating asymmetrical hybridisation between *M.fascicularis* and *M.mulatta*.

Thick white line: Approximate border between *M. fascicularis* and *M.mulatta*, according to Fooden.

Thin white line: Boundary of occurrence of infrazygomatic facial crests in *M. fascicularis*

X = Localities where specimens with intermediate tail length are known (Fooden, 1997)

geographic barrier. In effect, *M. fascicularis* differentiated as a rainforest species, while proto-*mulatta /cyclopis / fuscata*, north of the Isthmus of Kra, was adapted to deciduous forest.

- When the Isthmus of Kra became dry land, perhaps even as late as the end of the Pleistocene, the adaptable *M. fascicularis* spread north across it.

- It proved competitively superior to *M. mulatta* in the Semi-evergreen Forests that now dominated the Indochinese subregion, and spread north, interbreeding with *M. mulatta* as it went.

- Males of *M. mulatta* are larger than those of *M. fascicularis*, and presumably dominant to them. Interbreeding might then be predominantly unidirectional.

- *M. fascicularis* has gradually replaced *M. mulatta*, along a northward-moving frontier of hybridisation in which some *M. mulatta* males interbred with *M. fascicularis* females. The common occurrence in mainland *M. fascicularis* of relatively short tails, as well as of the infrazygomatic form of facial crest typical of *M. mulatta* (Fooden, 1995), are a legacy of this replacement with hybridisation. A *M. mulatta* Y chromosome was found in a monkey otherwise resembling *M. fascicularis* in Vietnam (Tosi *et al.*, 2000).

Macaca sinica group: *Macaca sinica, radiata, assamensis* and perhaps *pelops, thibetana, munzala.* The distribution is discontinuous: Sri Lanka and southern India, then a wide hiatus, and a continuous range from Nepal to northern Indochina and north into South-Central China. It is closely related to the *fascicularis* group, from which it would have separated somewhat over 3½ million years ago, and the southern (*sinica/radiata*) and northern (*assamensis/munzala/thibetana*) sections separated not long afterwards (Morales and Melnick, 1998). To explain both this distributional gap and the restriction of the group to just the northern parts of mainland Southeast Asia, I propose the following scenario:

- The *M. sinica* group was formerly spread throughout the Indian and Indochinese subregions.
- *M. mulatta* is the dominant species of Deciduous and Semideciduous Forest; as discussed earlier, it evolved in mainland East Asia north of what was then the Kra seaway. In the Middle Pliocene, it spread into India, splitting the range of the *M. sinica* group into two halves: a cool-climate subgroup in the sub-Himalayan area, South China and mainland Southeast Asia, and a subtropical group in southern India and Sri Lanka.
- The cool-climate subgroup then split into two proto-species: *M. thibetana/munzala* in the colder (mostly high altitude) areas, and *M. assamensis* adapted to the margins of these areas; *M. assamensis* in turn spread, restricting *M. thibetana* to South China, and leaving another high altitude isolate, the recently described *M. munzala*, in Arunachal Pradesh.

Macaca arctoides group: this contains *Macaca arctoides* alone. It is found, apparently discontinuously, throughout mainland Southeast Asia north of the Isthmus of Kra.

In what is one of the most stunning and unexpected results of recent DNA analyses, Tosi *et al.* (2000) showed fairly conclusively that *M. arctoides* is the result of an ancient hybridisation between female proto-*fascicularis* and male proto-*assamensis/thibetana*; this would have occurred, if the dating of the mitochondrial DNA split by Morales and Melnick (1998) is correct, some 1.6 million years ago. I hypothesise that, in a period when *M. mulatta* was restricted to China and/or India and representatives of the *M. sinica* group were living throughout mainland Southeast Asia, a population of *M. fascicularis* crossed the

Kra seaway and met and hybridised with proto-*assamensis/thibetana*. The hybrid population was successful, and spread north, replacing its paternal species except in the northern parts of the region (southern China, northern Burma, northern Indochina), and going on to develop its own unique and bizarre features. Later, perhaps in the Middle Pleistocene, *M. mulatta* spread into mainland Southeast Asia and proved capable of existing in sympatry with *M. arctoides*.

And what of *M.sylvanus*? The group includes a single species, *Macaca sylvanus*. Confined to North Africa (and, of course, Gibraltar), although during the Pleistocene it was fairly widespread in Western Europe. It differs strongly from other taxa in a number of features:

It is the only species of macaque in which the ischial callosities meet in the midline in the adult male, in which it resembles baboons and mangabeys - hence this condition is plesiomorphic. In most other genera of the Papionini, the males' ischial callosities actually fuse across midline; observations of adult male macaques on Gibraltar, however, show a condition in which they meet, but do not actually fuse.

It is the only macaque species, most of the Sulawesi species apart, with a distinctive skull - the other species being hard to differentiate cranially. Verheyen (1962) showed *M. sylvanus* has a wider interorbital pillar than other species of the genus, and I concur, from my examination of thee Natural History Museum (London) specimens (Figure 3). In *M. sylvanus* the interorbital pillar is 7.5-8.5mm wide, whereas in other species it is nearly always <7.5mm. In this respect however the species is not plesiomorphic, considering that all other members of the Cercopithecinae have a narrow pillar; the resemblance of *M. sylvanus* to the Colobinae must therefore be convergent.

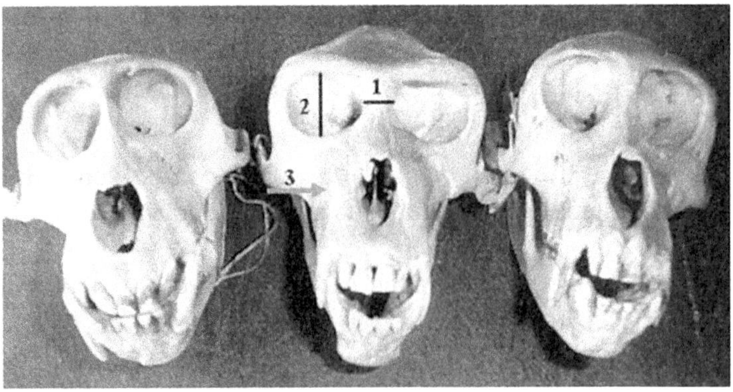

Figure 3. Distinctive features of the skull of Barbary macaques (centre) compared to other macaques (*M.nemestrina*, left, *M. assamensis*, right)

1. Inter-orbital pillar (relatively wide)
2. Orbits (squarer, smaller in vertical diameter)
3. Nasal ridges (present)

- According to my examination of the series in the NHM, *M. sylvanus* is distinguished by its low square orbits. The maximum vertical diameter of the orbit in *M. sylvanus* is 21-24mm; in other species it is rarely below 26mm.
- There is a trace of ridges on either side of nasal region, giving a "channeled" appearance. This is not seen in other species, except for some of the Sulawesi species.
- *M.sylvanus* has molar flare. This is a second plesiomorphic condition, shared with baboons and mangabeys. Other macaque species lack molar flare.

The mtDNA evidence (Morales & Melnick, 1998), confirms the deductions from morphology that *M. sylvanus* is the sister species to other members of the genus *Macaca*, corroborating hypotheses of an African origin for the genus (Abegg, this volume).

Postscript

The first time I visited Paris to study in the Muséum National d'Histoire Naturelle (long, long ago) I walked along the Boulevard St Germain, and lo! - a cafe called *Les Deux Magots*. Risking bankruptcy, I sat down at a table on the pavement outside, and ordered coffee. Before leaving, I peered inside. My hopes of seeing photographs or paintings of Barbary macaques were dashed. Instead, there were statues of two real *magots* – Chinese sages. But I did wonder whether, when John-Paul Sartre sat down, perhaps at the very same table, to discuss the arts and philosophy with Simone de Beauvoir and Giraudoux, the thought had crossed his mind that another kind of *magot* was perhaps not so far removed, in an existentialist sense, from his own life.

References cited

Abegg, C. & B.Thierry. 2002a. The phylogenetic status of Siberut macaques: hints from the bared-teeth display. *Primate report*, **63:** 73-78.

Abegg, C. & B. Thierry, B. 2002b. Macaque evolution and dispersal in insular south-east Asia. *Biol. J. Linn. Soc.* **75:** 555-576.

Bailey, J.F., M. Henneberg, I.B.Colson, A.Ciarallo, R.E.M. Hedges & B. Sykes. 1999. Monkey business in Pompeii – unique find of a juvenile Barbary macaque skeleton in Pompeii identified using osteology and ancient DNA techniques. *Mol.Biol.Evol.* **16:**1410-1414.

Cronin, J.E., R.Cann & V.M.Sarich. 1980. Molecular evolution and systematics of the genus *Macaca*. Pp.31-51 in Lindbergh, G.G., ed.,

The Macaques: Studies in Ecology, Behaviour and Evolution. New York: Van Nostrand Reinhold.

Eudey, A.A. 1980. Pleistocene glacial phenomena and the evolution of Asian macaques. Pp.52-83 in Lindbergh, G.G., ed., *The Macaques: Studies in Ecology, Behaviour and Evolution.* New York: Van Nostrand Reinhold.

Evans B.J., J.C.Morales, J.Supriatna & D.J. Melnick. 1999. Origin of the Sulawesi macaques (Cercopithecidae: *Macaca*) as suggested by mitochondrial DNA phylogeny. *Biol.J.Linnean Society*, **66:**539-560.

Fooden, J. 1976. Provisional classification and key to living species of macaques (primates: *Macaca*). *Folia primat.*, **25:**225-236.

Fooden, J. 1995. Systematic review of Southeast Asian Longtail Macaques, *Macaca fascicularis* (Raffles, [1821]). *Fieldiana, Zool.,* N. S., no. 81.

Fooden, J. 2000. Systematic review of the rhesus macaque, *Macaca mulatta* (Zimmermann, 1780). *Fieldiana, Zool.,* N. S., no. 96.

Goudsmit, J. & D. Brandon-Jones. 2000. Mummies of olive baboons and Barbary macaques in the Baboon Catacomb of the Sacred Animal Necropolis at North Saqqara. *J.Egypt.Archaeol.* **85:**45-53.

Groves, C.P. 1989. *A theory of human and primate evolution.* Oxford: Oxford University Press.

Groves, C.P. 2001. *Primate taxonomy.* Washington: Smithsonian Institution Press.

Jolly, C.J. & P.J. Ucko. 1968. The riddle of the Sphinx-monkey. Pp. 319-335 in M. Douglas and P. Kaberry, eds., *Man in Africa.* London: Tavistock Press.

Kitchener, A.C. & C.P. Groves. 2002. New insights into the taxonomy of *Macaca pagensis* of the Mentawai Islands, Sumatra. Mammalia, **66:**533-542.

Koppe, T. & Y. Ohkawa. (1999) Pneumatization of the facial skeleton in catarrhine primates. Pp. 77-120 in T.Koppe, H.Nagai & K.W.Alt, eds., *The Paranasal Sinuses of Higher Primates*, (Quintessence, Berlin),

Lacépède, B.G.E. de. 1799. *Tableau des divisions, sous-divisions, ordres et genres des mammifères.* Paris: Plassan.

Lynn, C.J. 1997. Excavations at Navan Fort 1961-71, County Armagh (original editor D. M. Waterman). *Northern Ireland Archaeological Monographs* No.3. Belfast: Stationery Office.

Malbrant, R. & A.Maclatchy. 1949. *Faune de l'Équateur Africain Français. II, Mammifères.* Paris: Paul Lechevalier.

Morales, J.C. & D.J.Melnick. 1998. Phylogenetic relationships of the macaques (Cercopithecidae: *Macaca*), as revealed by high-resolution restriction site mapping of mitochondrial ribosomal genes. *J.Hum.Evol.* **34:**1-23.

Roos, C., T.Ziegler, J.K.Hodges, H.Zischler & C.Abegg. 2003. Molecular phylogeny of Mentawai macaques: taxonomic and biogeographic implications. *Mol. Phyl. Evol.* **29:** 139-150.

Rosenblum, L.L., J.Supriatna & D.J.Melnick. 1997. Phylogeographic analysis of pigtailed macaque populations (*Macaca nemestrina*) inferred from mitochondrial DNA. *Amer.J.Phys.Anthrop.* **104:**35-45.

Tosi, A.J., J.C.Morales & D.J.Melnick. 2000. Comparison of Y chromosome and mtDNA phylogenies leads to unique inferences of macaque evolutionary history. *Mol.Phyl.Evol.* **17:**133-144.

Verheyen, W.N. 1962, Contribution à la craniologie comparée des primates: les genres *Colobus* Illiger 1811 et *Cercopithecus* Linné 1758. *Ann.Mus.Roy.Afr.Cent.*, ser. in 8°, **105**:1-211 + annexes, plates.

Woodruff, D.S. 2003. Neogene marine transgressions, palaeogeography and biogeographic transitions on the Thai-Malay peninsula. *J.Biogeog.* **30:**551-567.

The role of contingency in the evolution of the Barbary macaque

C Abegg
Department of Reproductive Biology, German Primate Centre, Kellnerweg 4, 37077 Göttingen, Germany

Macaques almost certainly originated in Africa around 7 million years ago. Since then, not only did the ancestor of the present-day Barbary macaque successfully populate Europe for several million years, but it also colonised broad areas throughout Asia. In this chapter, I will examine how historical contingency, i.e. the occurrence of chance events in the course of history (Abegg, 2004), affected the evolution of the earliest radiation of macaques over the African, European and Asian continents. After having proposed an evolutionary scenario for the deployment of early macaques, I will tentatively reconstruct the morphological features and habitat preferences of the common ancestor of all macaques. In order to explain the present distribution of Barbary macaques, I will attempt to show that contingent factors, including more recent influences linked to human pressure, have played a determining role.

The African origins and dispersal to Europe

Macaques are thought to have emerged in Africa from the *Papionin* tribe during the Miocene epoch (20 to 5 MYA) (Delson, 1980). The most likely origin of the Genus is North Africa, as indicated by 7 million-year-old fossils found in Algeria (in Marceau and named *M. flandrini*). Other remains of similar age and referred to as macaques have also been found in Central Africa (in Ongoliba, Congo). Delson (1975) suggests that both sets of fossils could represent early *Papionin* species which may have been distributed across what later became the Sahara desert, before the differentiation of the macaque Genus. Delson (1980) stated that "at the end of the Miocene, the Mediterranean almost completely dried up as a result of tectonic movements, and that this desiccation might have aggravated the general trend toward aridity already evident in southern Europe earlier in the late Miocene, eventually leading to the formation of a semi-desertic barrier across the Sahara." According to Delson, such a barrier could have led to the differentiation of the *Papionini* into the three currently existing groups (*Theropithecus*; *Papio* and *Cercocebus* south of the Sahara; and *Macaca* to the north). As the region of the Sahara became more arid during the Miocene, it could have constituted a barrier that prevented migration of primates between Central Africa and the North of the continent.

Macaca lybica (Stromer 1920), represented by a collection of dental remains found in Northern Egypt may, according to Delson (1980), predate (6 MYA) the division into Mediterranean and Asian lineages of the genus. By 6 MYA macaques already existed in North Africa, but precisely since when remains unclear. Overall however, the fossil records point to a North African origin of the macaque Genus.

Two main routes of dispersal could have then been used by an early form of macaque to spread out of Africa: one directly into Europe through a land connection across the Straits of Gibraltar and the other through the Middle-East, providing opportunities for dispersal into both Europe and Asia (Figure 1).

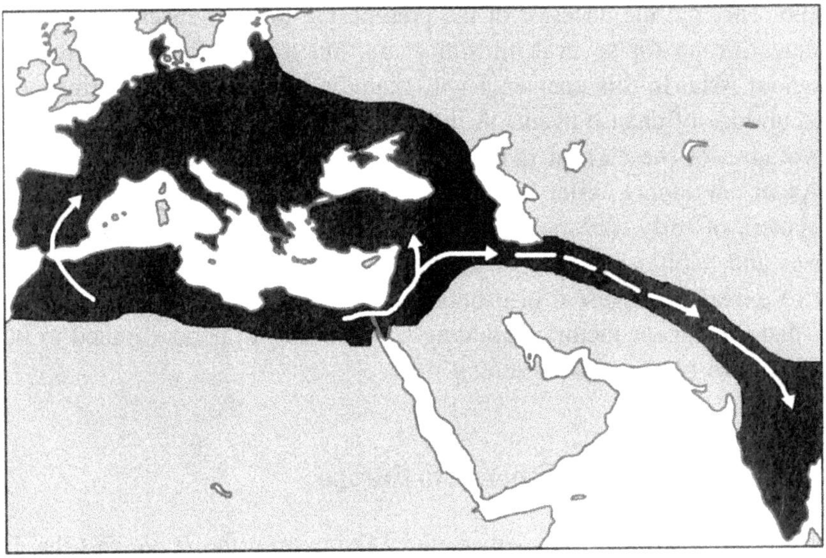

Figure 1. Macaque dispersal routes out of Africa

The arrows indicate one early possible dispersal route to Europe across Gibraltar, the main probable route to Europe along the east Mediterranean coast and a proposed route through the Middle-East to Asia.

The dotted lines show the possible location where dispersal to Asia was interrupted around 5 MYA.

The oldest remains of macaques in Europe are fossils of *M. sylvanus prisca*, the type-locality of which is near Montpellier, and which have been found in numerous localities in Southern and Central Europe (Italy, France, Germany, Hungary). Ranging in age from 5 to 3 MYA, they document the first extension of macaques through a crossing or circuit around the Mediterranean Sea into Europe. *M. sylvanus florentina* refers to fossils found in fields dating from the Late Pliocene to the Late Pleistocene (3 to 0.5 MYA) in Italy, but also in Spain, France, the Netherlands and Yugoslavia. These remains are most similar to present *M. sylvanus* while the older ones, *M. s. prisca*, are smaller. Fossils referable to *M. s. florentina* have been recovered dating back to the Middle Pleistocene (1 to 0.125 MYA) in present-day England, Spain, France, Germany, Italy, Central Europe, the Caucasus and Israel. These fossils were associated with inter-glacial warm phases and it seems that macaques failed to find a refuge during the last glacial period in Europe (22 000 to 10 000 years ago) as no younger remains have been recovered from Europe.

From their origin in North Africa, about 7 MYA, up to their disappearance from Europe, perhaps as late as the last glacial age starting 22 000 years ago, the Barbary macaque progenitor was able to maintain populations in changing habitats ranging from subtropical to temperate forest or steppe-like environments. Looking at fossil locations and associated remains of the fauna, a moderate climate – i.e. temperate or subtropical -seems to have represented the main requisite for macaque expansion in Africa and Europe. Ecological changes associated with colder climatic phases seem especially ill-suited to macaque survival in Europe. As global temperatures have repeatedly dropped and recovered over the last 5 millions years, *sylvanus*-type macaques either reduced their population to refuge areas in the South of the continent (Figure 2C) or disappeared completely from Europe and survived only in North Africa (Figure 2D).

In Europe, forced retreat of macaques during glaciations probably resulted in a mixing of gene pools. Thus, climatic fluctuations resulted in genetic homogenization of populations and it is possible that this occurred not only within Europe but also between populations in Africa and Europe. Indeed, today's population of Barbary macaque in North Africa can be seen as a relict (Albrecht, 1978) and some *sylvanus*-type populations have most probably succeeded in continuously occupying North Africa since their origin over 5MYA. As the youngest European remains - *M. s. florentina* – are nearest to the present-day *M. sylvanus*, it could be that the older *M. s. prisca* first became extinct in Europe at the advent of an early glacial only to be then replaced by a successful coloniser stemming from a surviving population in North Africa when the climate recovered. The existence of a glacial period around 3 MYA, at the time fossils of *M. s. florentina* replace the ones of *M. s. prisca*, would favor such a view. This would mean that macaques succeeded at least once in re-colonizing Europe around the Mediterranean Sea, thus favoring homogenization of populations on its northern and southern shores. Past climate fluctuations and their consequences on habitat availability as well as the existence of a corridor for dispersal between Africa and Europe through the Middle-East would provide a reasonable explanation as to why macaque populations did not diversify into various species there, contrary to what happened in Asia.

From Africa to Asia

At the end of the Miocene (5 MYA), macaques expanded out of Africa through the Middle-East toward the Caucasus and Europe but also deeper into Asia (Figure 1). Macaque dispersal from Africa to Europe and Asia must have occurred at a time when forested landscapes were available along their way through the Middle East, which was probably the case during the Miocene and to a lesser extent during the Early Pliocene, but not later on. As climate during the Late Miocene became progressively drier and cooler, landscapes in North Africa, Europe and the Middle-East shifted toward more open vegetation. Asian macaque progenitors were probably early isolated from their African ancestors at the end of the Miocene when climate changes transformed more forests into open habitats, thereby interrupting their route of dispersal through the Middle East. It can thus be hypothesised that a forested corridor once allowed macaques to disperse into Asia and that it was interrupted, probably somewhere between Mesopotamia, Afghanistan and Pakistan (Figure 1). Genetic studies together with fossil records and paleo-climatology provide evidence that the timing of such an interruption of gene flow between *M. sylvanus's* progenitor and the ancestor of Asian macaques was around 5.5 MYA (Delson, 1996; Roos *et al.*, 2003).

Available reconstructions of past climate and landscape over the last 6 million years (e.g. Morley, 2000) indicate that, although moist forest predominated in Southeast Asia, this was not the case in North Africa, Europe or most of the Middle-East. Thus, from their origin in North Africa, as well as along their dispersal routes into Europe and Asia, ancestral macaques would have to have been able to cope with deciduous, seasonal, and temperate or subtropical landscapes, sometimes with a reduced forest cover. On this basis, one has to assume therefore that colonization and adaptation into the tropical and equatorial Asian type of habitats with a more continuous forest cover further east and south must have been secondary. Adaptation of macaques to rainforests probably occurred only once Asia was colonised.

First radiation within Asia: the role of contingency

After macaque populations in Africa and Asia became separated, the first lineage that diverged in Asia (around 5.5 MYA) gave rise to the *silenus* group. Today, this comprises *M. silenus*, *M. nemestrina*, *M. leonina*, *M. siberu*, *M. pagensis* and the 7 Sulawesi island species and has a wide but highly fragmented distribution over south and southeast Asia. For a long time this distribution pattern was interpreted as being primarily due to competitive exclusion (Fooden, 1982). According to this view, competition was the principal driving force, with climate changes and associated modification of sea levels and land masses accounting for the colonization of the peripheral island regions of Sulawesi and the Mentawai. In this way, the macaques of the Sulawesi and Mentawai islands are generally considered to have descended from a continental pigtailed macaque (i.e. a *nemestrina*-like ancestor), itself derived from a *silenus*-like ancestor, finally colonizing their respective islands by rafting and by means of a land-bridge (Fooden, 1975; Delson, 1980; Eudey, 1980; Evans *et al.*, 1999; Abegg and Thierry, 2002). Whilst a consensus exists concerning their method of colonization, whether or not they actually derived from continental populations of pigtailed and liontailed macaques remains a matter of debate.

In 2002, an alternative model was proposed by Abegg and Thierry in which climate changes were accorded a much more central role. Since it has already been shown that climatic cycles have induced repeated contractions in the distribution of those primate taxa that survived in available habitats, sometimes followed by expansions (Jablonski and Whitfort, 1999; Jablonski *et al.*, 2000), it is reasonable to assume that glacial periods responsible for the substantial sea level falls that allowed macaques to colonise the peripheral islands, also had profound influences on macaque distribution on the mainland. Thus, the question as to where the progenitors of the *silenus* lineage could have maintained populations in continental areas despite such an intense glacial period, remains open. Refuges likely to keep a wetter climate during glacial periods, thus retaining forest habitat and macaque populations, have been proposed (Eudey, 1980; Morley, 2000; Abegg and Thierry, 2002), although according to Brandon-Jones (1996) they might have completely disappeared from mainland Southeast Asia, surviving only on the Sulawesi and Mentawai islands.

Here, and in light of the most recent genetic results (Ziegler *et al.*, in press), I present an alternative dispersal scenario to explain the present distribution of the *silenus* lineage. After a broad dispersal of the *silenus* lineage's progenitor in Southeast Asia (Figure 2B), most populations were extirpated as a result of climatic change (Figure 2C), with the exception of two that were isolated respectively on Sulawesi and the Mentawai islands (Figure 2D), that acted as refuges for the lineage. Subsequently, some macaques rafted from Siberut to the Sumatran mainland and re-colonised continental areas where their ancestors had already become extinct (Figure 2E). This stock gave rise to the liontailed macaque, the Indochinese and Sundaic pigtailed macaques (Figure 2F).

The peripheral macaque populations on Sulawesi and the Mentawai islands would thus represent the first offshoots of the *silenus* lineage derived from the African stock colonizing Asia. They originated as they succeeded in finding refuge while their progenitors failed to survive in continental areas. That their ancestors were more likely to have been restricted to their island refuges as a result of climatic changes rather than being superseded by macaque competitors from new, expanding lineages, is a hypothesis that underscores the importance of contingency. Thus, an integration of contingency events

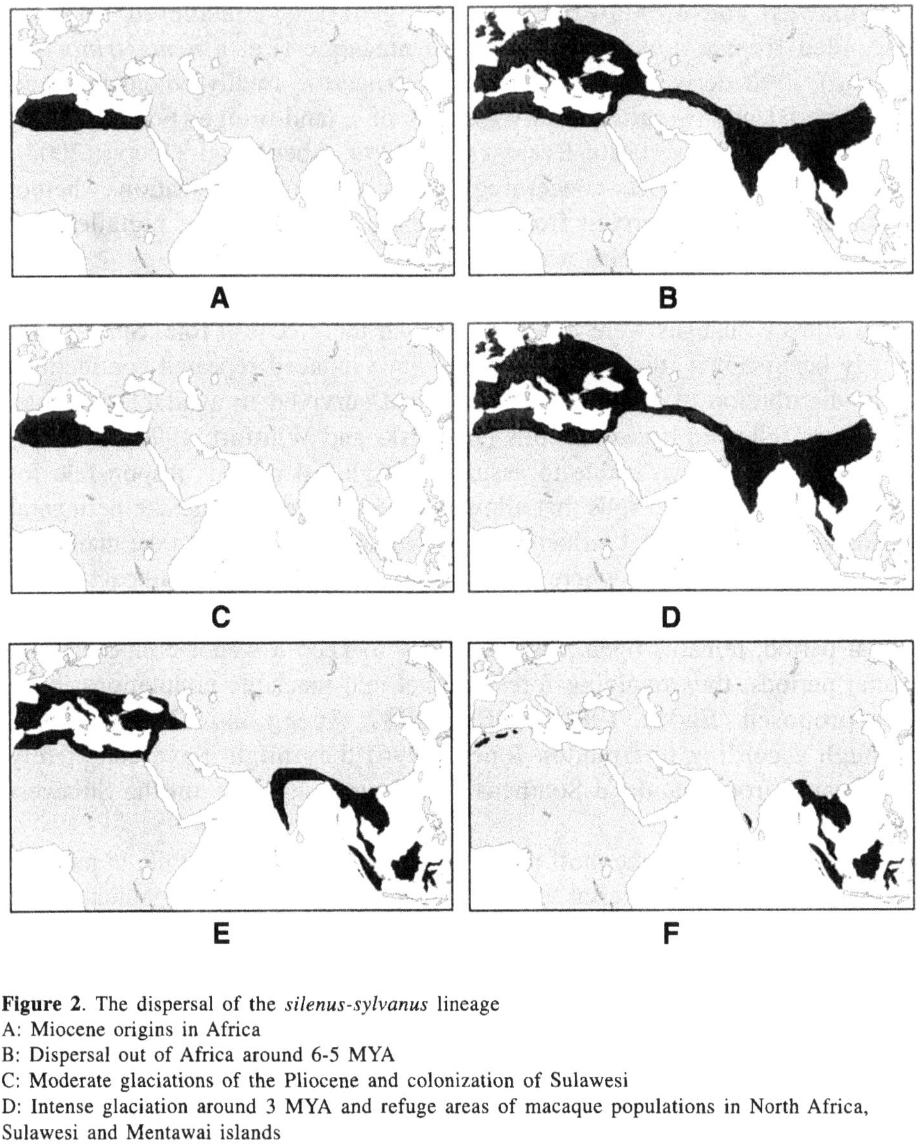

Figure 2. The dispersal of the *silenus-sylvanus* lineage
A: Miocene origins in Africa
B: Dispersal out of Africa around 6-5 MYA
C: Moderate glaciations of the Pliocene and colonization of Sulawesi
D: Intense glaciation around 3 MYA and refuge areas of macaque populations in North Africa, Sulawesi and Mentawai islands
E: Climate recovery and re-expansion of macaque distribution
F: Present distribution of the *silenus-sylvanus* lineage.

(climate and habitat changes, dispersal route interruption, emergence of a land bridge and occurrence of rafting) in the evolution of macaques, along with competitive exclusion and adaptation to changing habitats, accounts for the fragmented distribution of macaques, and their presence or absence in the islands of Southeast Asia.

Present distribution of the Barbary macaque

Contingency may have been an important factor influencing the evolution and distribution of early Asian macaques, but what of its role in influencing present day distribution of the Barbary macaque? The Barbary macaque progenitor, after its initial dispersal to Europe and Asia, was finally restricted to its present range because of climatic factors resulting in habitat changes from predominantly forested to drier and more open habitats. Around the Mediterranean Sea, as in Asia, climatic changes have strongly influenced the dispersal and distribution of macaque populations. The slopes of the Atlas Mountains in North Africa are thought to have represented one of the safest refuge areas in times of prolonged drought, along with some other smaller regions of Southern Europe. Over the last 5 million years, during warm and wet periods, the Barbary macaque was able to expand successfully in North Africa and Europe while forest was recovering. The Middle-East (more precisely the East Mediterranean coastland, see Figure 1) was most probably the route used by macaques to re-colonise Europe from North Africa when they disappeared from Europe.

The Barbary macaque would finally have disappeared from Europe during the last glacial, which ended some 10,000 years ago (Delson, 1980), probably having been held in check and then finally eliminated due to the combined influences of lack of suitable forest habitat and hunting pressure. Hunting as a contributing cause to macaque extinction in Europe is plausible for two reasons. Firstly, it has already been proposed that the contraction of Barbary macaque populations in wide areas of North Africa was partly due to hunting (Taub, 1984). Secondly, during the last glacial period, the warmer regions of Southern Europe probably acted as refuge areas for *Homo sapiens* (and possibly *Homo neanderthalis*) as well as for *Macaca* so that hunting pressure may have been coupled with reduced availability of suitable habitat. In North Africa, the present mountainous habitat distribution of the Barbary macaque can be considered to have resulted from degradation of lowland forests caused by humans rather than from adaptation to the particular types of habitats found at higher altitudes, where most remnant populations are now found. Contingent factors stemming from climatic changes and human pressure are thus proposed to have been determinant in shaping the present distribution of Barbary macaques and in leading to their extinction in Europe.

The ancestral form of macaque

The Barbary macaque is thought to represent the closest living relative of the ancestor of all macaques. Does this mean that the Barbary macaque also resembles most the ancestor of all macaques?

It has been alternatively postulated that the ancestor of both Barbary macaques and Asian macaques was either a long-tailed primate adapted to evergreen forest (Fooden, 1975; Delson, 1980), therefore most closely resembling *M. silenus*, or a short-tailed macaque adapted initially to seasonal habitats then secondarily (after reaching Asia) to evergreen forest and resembling the Barbary macaque (Eudey, 1980; Fa, 1989.

As we have seen, at the end of the Miocene, macaques started to venture out of Africa through the Middle-East toward the Caucasus and Europe but also more deeply into Asia. If we recognise the earliest Asian macaque taxon as the one leading to the Sulawesi, Mentawai and liontailed macaques (Fooden, 1980), then it follows that these species probably retained ancestral characteristics of the genus. These characteristics must also have been shared by the ancestor of *M. sylvanus*. Could modern Barbary macaques, with their short tail and pale fur colour, be considered as the living ancestor of all macaques even though most of the direct descendants of the first line of Asian macaques show quite different morphological features (e.g. *M. silenus* has a long tail and shares with Sulawesi and Mentawai macaques a dark pelage as well as a black skin)?

Fa (1984; 1989) proposed that the Barbary macaque retained the habitat preferences of its progenitor, originally adapted to seasonal climate but that, to the contrary, the progenitor of *M. silenus* became secondarily adapted to evergreen forest habitat. Although Fa did not overtly state that macaques living around the Mediterranean millions of years ago were already short-tailed, this can be considered implicit in his reasoning. Assuming that, during the habitat shifts brought about by climatic changes over millions of years, there has been a trend toward tail reduction in each macaque radiation (*sinica* and *fascicularis* lineages) as an adaptation to more terrestrial habits and/or colder climates, and that global temperatures cooled from the Pliocene up to the Pleistocene, it is logical that short tails predominate among extant living species (19/22). Interestingly, the only species with a long tail within the *silenus-sylvanus* lineage is the liontailed macaque. This suggests that only the macaque progenitors that became more arboreal instead of evolving toward semi-terrestrial habits gave rise to a species with a long tail. As *M. silenus* is derived from a common ancestor with Barbary, Sulawesi and Mentawai macaques, the descendants of which are all short-tailed, it might have evolved a long tail while adapting to arboreal habits in its rainforest refuge of South India.

Several assumptions can be made:

1 Barbary macaques' progenitors had a short tail in their African and European range.

2 Mentawai, Sulawesi and pigtailed macaques have retained the same morphological feature in their Southeast Asian range but not the liontailed macaque.

3 The liontailed macaque is derived from an ancestor with a black skin, dark pelage and short tail, which supports the hypothesis that refuges gave rise to mainland forms.

Interestingly, if we compare the morphology of all macaques from the first radiation, Barbary and pigtailed macaques are the only species not showing a dark pelage and a black skin. These species have also long maintained a wide continental distribution. They have done so up to present times despite alternation of glacial and warm periods. Indeed, pigtailed macaques are still found over a large part of mainland Southeast Asia and the Barbary macaque, before historic times, was able to expand and populate a broad area over Northern Africa and Europe. It seems that pale fur colors as well as short tails represent adaptive features for macaques forced to live in rather open landscapes. Within the *silenus-sylvanus* lineage, only representatives that have been repeatedly forced to cope with more open habitats during climatic fluctuations would have evolved pale fur and skin colors. It is thus proposed here that the ancestor of all macaque species was semi-terrestrial with a short tail and pale fur and whose descendants in Africa, Europe and Asia went through the repetitive ecological changes triggered during intensifying Pleistocene climatic oscillations.

Conclusion

The evolution of macaques has generally been considered from the standpoint of adaptation and competition between species (Fooden, 1976; 1980). For instance, the gradual replacement of ancestral populations through competition was held to be the main mechanism responsible for the disappearance of the first wave of macaques in Asia and its replacement by more recent colonisers. Numerous events, however, have occurred through geological time, creating ample opportunity for the occurrence of historical contingencies. In times of adverse climatic changes such as those associated with a global drop in temperatures, contingent events are more likely. Populations may, for example, be drastically disjoined and reduced due to habitat fragmentation; the existence of refuges, the emergence of land bridges and the possibility of sea rafting may then strongly affect the fate of animal populations (Haffer, 1969; Simpson, 1983).

The consequences of climate changes, which have markedly affected macaque dispersal and evolution over the last million years, are now being reinforced, though on a shorter time-scale, by the global consequences of human activities on natural landscapes. Habitat alteration, a contingent event for species, can today be observed on a broad scale. It may be that the rapid changes brought about by humans will not leave macaques the time and space they need to survive.

References

Abegg, C. and Thierry, B. 2002. Macaque evolution and dispersal in insular Southeast Asia. *Biological Journal of the Linnean Society* **75**: 555-576.

Abegg C. 2004. The role of contingency in evolution. In: *How Societies Arise: The Macaque Model*. Thierry, B., Singh, M. and Kaumanns, W (eds.), Cambridge University Press, Cambridge.

Albrecht GH. 1978. *The craniofacial morphology of the Sulawesi Macaques: multi-variate approaches to biological problems*. Basel: Karger.

Brandon-Jones D. 1996. The Asian Colobinae (Mammalia: Cercopithecidae) as indicators of Quaternary climatic changes. *Biological Journal of the Linnean Society* **59**: 327-350.

Delson E. 1975. Paleoecology and zoogeography of the old world monkeys. In R. Tuttle, ed. Primate functional morphology and Evolution. The Hague: Mouton, pp. 37-64.

Delson E. 1980. Fossil macaques, phyletic relationships and a scenario of deployment. In: Lindburg DG, ed. *The macaques. Studies in ecology, behavior, and evolution*. New York: van Nostrand Rheinhold, pp. 10-30.

Delson E. 1996. The oldest monkeys in Asia. In: *International Symposium: Evolution of Asian Primates*, Freude and Kyoto University Primate Research Institute, Inuyama, Aichi, Japan, p.40.

Eudey AA. 1980. Pleistocene glacial phenomena and the evolution of Asian macaques. In: Lindburg DG, ed. *The Macaques. Studies in ecology, behavior, and evolution*. New York: van Nostrand Rheinhold, 52–83.

Evans BJ, Morales JC, Supriatna J, Melnick DJ. 1999. Origin of the Sulawesi macaque as suggested by mitochondrial DNA. *Biological Journal of the Linnean Society* **66:** 539–560.

Fa JE. 1984. Habitat distribution and preference in Barbary macaques (Macaca sylvanus). *International Journal of Primatology* **5**: 273-286.

Fa JE. 1989. The genus *Macaca*: a review of taxonomy and evolution. *Mammal Review* **19:** 45–81.

Fooden J. 1975. *Taxonomy and evolution of liontail and pigtail macaques (Primates: Cercopithecidae*). *Fieldiana: Zoology* **67**: 1-168.

Fooden J. 1976. Provisional classification and key to living species of macaques (Primates: *Macaca*). *Folia Primatologica* **25:** 225–236.

Fooden J. 1980. Classification and distribution of living macaques (*Macaca* Lacépède, 1799). In: Lindburg DG, ed. *The macaques. Studies in ecology, behavior and evolution.* New York: van Nostrand Rheinhold, 1–9.

Fooden J. 1982. Ecogeographic segregation of macaque species. *Primates* **23**: 574-579.

Haffer J. 1969. Speciation in Amazonian forest birds. *Science* **165**: 131-137.

Jablonski NG, Whitfort MJ. 1999. Environmental changes during the Quaternary in East Asia and its consequences for mammals. *Records of the Western Australian Museum (supplement)* **57:** 307–315.

Jablonski NG, Whitfort M, Roberts-Smith N, Qinqi X. 2000. The influence of life history and diet on the distribution of catarrhine primates during the Pleistocene in East Asia. *Journal of Human Evolution* **39**: 131-157.

Morley RJ. 2000. *Origin and evolution of tropical rain forest.* Chichester NY: Wiley.

Purvis A. 1995. A composite estimate of primate phylogeny. *Philosophical Transactions of the Royal Society, London* **348**: 405-421.

Roos, C., Ziegler, T., Hodges, K., Zischler, H. and C. Abegg. 2003. Molecular phylogeny of the Mentawai macaques: taxonomic and biogeographic implications. *Molecular Genetics and Evolution* **29**:139-150.

Simpson GG. 1983. *Fossils and the History of Life.* New York: Scientific American Library.

Taub, DM. 1984. A brief historical account of the recent decline in geographic distribution of the Barbary macaque in North Africa. In Fa ed. *The Barbary Macaque – a Case Study in Conservation.* Plenum Press, New York.

Barbary but not barbarian: social relations in a tolerant macaque

B Thierry[1] and F Aureli[2]
[1]*Departement d'Ecologie, Physiologie et Ethologie, Centre National pour la Recherche Scientifique, UMR 7178, 23 rue Becquerel, 67087 Strasbourg, France*
[2]*Research Centre in Evolutionary Anthropology and Palaeoecology, School of Biological and Earth Sciences, Liverpool John Moores University, James Parsons Building, Byrom Street, Liverpool L3 3AF, UK*

Our views about macaque societies have long been shaped by the study of rhesus and Japanese macaques (*Macaca mulatta*, *M. fuscata*). It was a historical accident that research about the social behaviour of these species developed earlier and more extensively than for any other monkey, and that both species exhibit similar social relations, including weak interindividual tolerance, intense aggression and submission, strong nepotism, strict hierarchies, and little affiliation between mature males (Thierry *et al.*, 2004). Such patterns were believed to be typical of the genus *Macaca* until the seventies. Then, our knowledge increased regarding other macaque species and first of all the Barbary macaques (*M. sylvanus*). The accumulation of information during the last decades showed that the behaviour of Barbary macaques significantly departs from the above picture. Whereas this conferred more depth to our vision of the social organisation of macaques, it also raised questions about the evolutionary significance of such variations. In what follows we successively review the patterns of affiliation and conflict resolution of Barbary macaques, their social relationships and networks, then the conditions of mating competition, in order to point at the significance of the links between the various characters of the species' social organisation.

Patterns of affiliation and conflict resolution

Macaque species widely differ in the degree of asymmetry of contests. Whereas the target of aggression generally flees or submits in rhesus and

---

This chapter is dedicated to Cédric Marengo, promising young scientist, lover and student of Barbary macaques, who suddenly died the 9th of November 2002 from aneurysmal rupture.

Japanese macaques, s/he protests or retaliates in two thirds of conflicts in macaques originating from Sulawesi island (*e.g. M. tonkeana, M. nigra*) (Thierry, 2000). Rates of reconciliation – *i.e.* a form of conflict resolution in which former opponents exchange affiliative behaviour soon after an aggressive conflict (de Waal and van Roosmalen, 1979; Aureli and de Waal, 2000) – also covary with the degree of conflict asymmetry: asymmetric relationships may inhibit the occurrence of affiliative contacts between opponents, whereas conciliatory behaviour may facilitate information exchange and reduce social tension between more evenly matched individuals. Conciliatory tendencies are assessed from the frequency of post-conflict affiliative contacts between previous opponents while controlling for baseline levels of affiliation (*cf.* Veenema *et al.*, 1994). Conciliatory tendencies between unrelated individuals consistently range between 4 and 12% in rhesus and Japanese macaques while they are about 50% in Sulawesi macaques (Thierry, 2000).

The percentages of counter-aggression – a measure of conflict asymmetry – in agonistic interactions between unrelated individuals are relatively elevated in a semi-free ranging group of Barbary macaques (Kintzheim, France), the target countering about half of threats and attacks (Figure 1). In the same group, conciliatory tendencies ranged between 10 and 22% depending on age-sex categories (Figure 1), and post-conflict contacts between opponents were often accompanied by an affiliative facial expression (79% of events) and clasping or embracing (53% of events) (Marengo, unpublished data). The conciliatory tendency for dyadic conflicts between unrelated females in another captive group (Apenheul, The Netherlands) was 28% (*cf.* Aureli *et al.*, 1994, 1997). Based on these values, Barbary macaques appear intermediate in dominance style (*sensu* de Waal and Luttrell, 1989) among macaque species. Macaques have been arranged on a 4-grade scale based on their affiliative and agonistic behaviours, from the least tolerant species to the most tolerant ones (Thierry, 2000). Rhesus and Japanese macaques are set in grade 1, which is characterised by mainly unidirectional conflicts and reduced conciliatory tendencies, whereas Sulawesi macaques are placed in grade 4, which is characterised by high levels of counter-aggression and conciliatory tendencies. From their patterns of conflict and reconciliation, Barbary macaques are assigned to grade 3, meaning that they belong to the tolerant side of the scale.

Conflict resolution after aggression may not only include friendly behaviour such as reconciliation. Redirection of aggression, which consists of the victim attacking a third individual soon after the initial conflict, has been reported in several macaque species (Scucchi *et al.*, 1988; Aureli and van Schaik, 1991a). It appears to reduce tension and the likelihood of renewed aggression (Aureli and van Schaik, 1991b), and could also signal to third parties the victim's post-conflict condition and motivation to pre-empt subsequent challenges by them (Kazem and Aureli, 2005). Barbary macaques were found to display redirection of aggression when post-conflict behaviour was compared with control situations (Aureli *et al.*, 1994). One variation of this post-conflict behaviour is kin-oriented redirection in which the victim of the initial aggression redirects against a former aggressor's vulnerable kin. Kin-oriented redirection has been investigated and found in Japanese macaques in low-risk situations (Aureli *et al.*, 1992), and there is some evidence for pigtailed (*M. nemestrina*) (Judge, 1982) and longtailed macaques (*M. fascicularis*) (Aureli and van Schaik, 1991a). There has been no study on kin-oriented redirection in Barbary macaques, but it is possible that it does not occur in this and other tolerant species because victims may not need to wait for suitable opportunities and retaliate against opponents' vulnerable kin as they can directly counter-attack the aggressor (*cf.* Thierry, 1990).

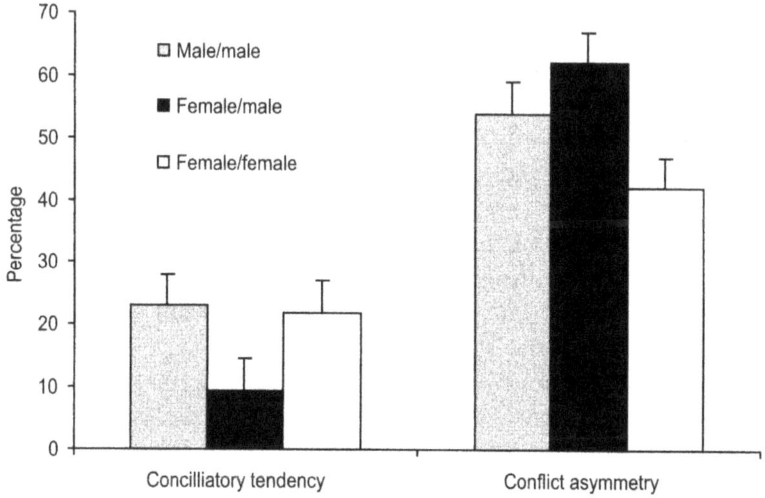

Figure 1. Conflict asymmetry and conciliatory tendency among adult Barbary macaques (Kintzheim, France). Conflict symmetry is measured as the percentage of conflicts (means + SEM) in which the target of aggression protests or counter-attacks; it is calculated by pooling all conflicts having occurred between mature individuals. Conciliatory tendencies are calculated according to Veenema *et al.* (1994) from dyadic conflicts occurred between pairs of unrelated individuals. N = 24 males at least 5-yr old and 40 females at least 4 yr-old (Marengo, unpublished data).

The use of specific behaviours, aiming to reduce social tension and facilitate friendly contacts, is a typical trait of species from grades 3 and 4. Peaceful intervention in conflicts represents a first type of appeasement used by Barbary macaques at the triadic level. When a conflict arises between two parties, a third individual may approach and clasp, mount or groom one of the opponents while emitting affiliative facial expressions like lipsmacking or teeth-chattering. The intervener may so protect the beneficiary without endangering his relationship with the recipient contrary to what happens in aggressive intervention. Peaceful interventions are rarely reported among non-human primates. In the semi-free ranging population of Kintzheim, they occurred in as many as 10% of conflicts (N = 1268 conflicts) (Marengo, 2002), a percentage similar to those reported in Sulawesi macaques (Petit and Thierry, 1994, 2000). In Barbary macaques, peaceful interventions stop aggression more frequently than aggressive interventions within 5 s (56% vs. 42 % of conflicts respectively: $P < .05$, $z = 2.0$, N = 13 interveners, Wilcoxon test) (Marengo, 2002).

Infants play a role in the mediation of social interactions in Barbary macaques. The great amount of attention and handling devoted by males to infants has struck observers from the start of research on Barbary macaques (Lahiri and Southwick, 1966; Deag and Crook, 1971). Subadult and adult males regularly pick, carry, huddle, groom, defend or play with infants of both sexes (Figure 2a) (Deag, 1980; Taub, 1984; Paul *et al.*, 1996; Ménard *et al.*, 2001). Males also use infants in the so-called 'triadic male-infant interactions'. In such interactions, either a male approaches another male who holds an infant or a male carries an infant while approaching another male. Then, the two males direct affiliative signals toward each other and the infant, holding the infant and sometimes lifting her up; the role played by the infant is typically passive (Figure 2b) (Deag, 1980). The selectivity of males is a characteristic feature of their behaviour. Each male specifically cares for certain infants while ignoring others, and he chooses these infant partners when engaging in triadic interactions (Deag, 1980; Taub, 1984; Paul *et al.*, 1996; Ménard *et al.*, 2001).

The quest for the reasons of males' investment in infants has consistently marked the study of Barbary macaques. According to the 'agonistic buffering hypothesis', males use infants as social tools to inhibit aggression from others and regulate their relationships (Deag and Crook, 1971; Deag, 1980). In its early version, the hypothesis specified that subordinate males use infants as social tools to reduce the likelihood of being attacked by higher-ranking individuals. In fact, triadic interactions are rarely associated with conflicts, and a dominant male may pick an infant to approach a lower-ranking partner, even if the reverse interaction is more frequent (Deag, 1980; Taub, 1980a; Smith and Peffer-Smith, 1982). Further hypotheses have been formulated. They emphasise infant caretaking, which can be either kinship investment if males preferentially associate with their own offspring or relatives (Taub, 1980a, 1984), or mating effort if caretakers increase this way their chance of being chosen by the infant's mother as a mating partner (Ménard *et al.*, 2001). Although infant care may be associated with higher mating frequencies with the mother in the next reproductive season, paternity analyses have shown that there is no relation between the probability of fathering by males and

Figure 2. Social behaviours in semi-free ranging Barbary macaques (Kintzheim, France). (a) Adult male carrying an infant; (b) Triadic male-infant interaction; (c) Triadic female-infant interaction; (d) Young female emitting a silent bared-teeth display. Photographs by B. Thierry.

their caretaking activities both in captive and wild populations (Paul *et al.*, 1996; Ménard *et al.*, 2001). At present, agonistic buffering appears as the better-supported explanation for the occurrence of triadic male-infant interactions, which may not only have the short-term function of inhibiting aggression. Whereas male carriers often have a caretaking relationship with the infant used in an interaction, it is remarkable that caretakers are also the preferred recipients of the triadic interactions initiated by males carrying the infant, indicating that the latter exploit the familiarity between infant and recipients (Paul *et al.*, 1996). By promoting affiliative interactions among males, the use of infants may bear a long-term function, namely increasing social tolerance among males and establishing strong bonds between them.

It may not be coincidental that the occurrence of triadic male-infant interactions is associated with exceptionally high rates of infant handling in

Barbary macaques. A similar association is reported in Tibetan macaques (*M. thibetana*) (Deng, 1993; Ogawa, 1995; Zhao, 1996). It has been argued that triadic interactions are "a specialised ritualised subset of a comprehensive system of male-caretaking" (Taub, 1980a, p. 196), in which case infant use may derive from infant care (Zhao, 1996). Conversely, if the agonistic buffering hypothesis holds true the strong attraction of males toward infants would represent a prerequisite for their strategies for relationship negotiation between males (Paul *et al.*, 1996; Thierry, 2000). They are bound to invest time and establish close relationships with infants in order to successfully use them in their interactions.

Although the emphasis for triadic interaction with infants has been on males, we must add that females are not absent from the picture. Mothers are quite permissive, they leave their infant interacting with males and other females. Such permissiveness is a characteristic feature of the most tolerant species on the macaque continuum (Maestripieri, 1994; Thierry, 2004). Actually individuals of all age-sex categories display high rates of positive infant-handling behaviour (Small, 1990a; Paul and Kuester, 1996). Triadic interactions are common among females too (Figure 2c) with the caveat that they are usually approached when they hold an infant rather than carrying an infant to another female (Smith and Peffer-Smith, 1982; Small, 1990a; Paul and Kuester, 1996; Timme, 1996). Following triadic interactions, partners frequently stay in contact, resting or grooming each other. Like males, females may use such interactions to strengthen their social relationships.

Social relationships and networks

As for other macaques, dominance and kinship relationships structure the social organisation of Barbary macaques. The species exhibits clear-cut hierarchies and females form matrilines, i.e. subgroups of maternal relatives who help each other in contests (Deag, 1977; Taub, 1980a, b; Paul and Kuester, 1987; Paul, 1989; Widdig, 2000). Nonetheless the asymmetry of their dominance relations and their degree of nepotism are far from being extreme as those described in rhesus and Japanese macaques for instance. A relevant marker of the dominance style between conspecifics in macaques is the silent bared-teeth display (Preuschoft and van Schaik, 2000; Thierry, 2000). When performing this facial expression, the individual retracts the lips and exposes the teeth (Figure 1d). In species from grade 1 this is a subordination signal: the sender submits at the approach of a higher-ranking individual. In contrast, in species from grade 4 the same display does not have a communicative function about dominance status, it signals the sender's peaceful intentions. In Barbary macaques the silent-bared-teeth display may bear both meanings.

Depending on the behavioural context it may alternatively express subordination or lead to an affiliative interaction (Preuschoft, 1992). It is also a frequent answer to the teeth-chattering display, another facial expression characterised by retraction of the lips and repeated moving up and down of the lower jaw. Teeth-chattering is exclusively reported in macaques from grades 2 and 3. In Barbary macaques it conveys a positive meaning; for example, it commonly occurs in the context of triadic interactions around infants (Deag, 1980; Preuschoft, 1992).

Sex differences have been found in competition tests where the higher-ranking individual in a pair gets the incentives (Preuschoft *et al.*, 1998). When a peanut is at stake, females generally threaten their rival to get the food. In the same situation, males often refrain from overt competitive behaviour; in a majority of cases none of them shows any assertiveness over the other. When males are competing for a female in oestrus, however, they are more prone to recruit allies. It is likely that the readiness of individuals to enter into competition depends on the value of the resource and the risk incurred in obtaining it. For males in particular, well-armed rivals of equivalent power are at risk of conflict escalation. These results point at a relaxed dominance style in Barbary macaques (Preuschoft *et al.*, 1998). The appeasement behaviours regularly occurring between males – *e.g.* triadic male-infant interactions – may decrease the intensity of open contests or even prevent them. Moreover, in the mating season a majority of the conflicts occurring among adult males involves a counter-aggression (Kuester and Paul, 1992). In spite of the different competitive tactics of males and females, high rates of bidirectional agonistic interactions are observed in both sexes, confirming that the level of rank asymmetry is relatively weak between group members (Figure 1).

Barbary macaques additionally deviate from rhesus and Japanese macaques on another pattern, the "youngest ascendancy rule" (Chapais, 2004). In the latter species, females outrank their older sisters owing to the support of their mother and relatives in conflicts. This typically produces an age-inversed hierarchy among adult sisters, the youngest dominating the elders. In Barbary macaques, females inherit their mother's rank as in other macaques, but most females remain subordinate to their older sisters (Paul and Kuester, 1987; Prud'homme and Chapais, 1993). They appear unable to outrank them because of insufficient support both from non-kin and kin females. Furthermore, it seems that females may improve their dominance status by developing alliances with unrelated dominant females, grooming them or attending their infant. Two cases have been reported in which a female significantly rose in rank owing to the support of high-ranking females (Small, 1990b; Paul and Kuester, 1996).

Nepotism and dominance are linked through the occurrence of alliances

(Thierry, 2000). When the bias in favour of kin is only slightly pronounced, close ties exist even between non-relatives, coalitions between them are regular and dominance asymmetry is consequently weaker. The proportion of aggressive support provided by non-kin is especially high in Barbary macaques relative to other species (Aureli *et al.*, 1997). The openness of female social networks is further revealed by the analysis of partner preferences. When comparing the grooming networks of mothers and adult daughters, no shared partner preferences were found in Barbary macaques (Schino *et al.*, 2004). The investigation of kinship effects on reconciliation rates further pointed at the relatively weak nepotism of Barbary macaques: the measure of conciliatory tendencies between kin and between non-kin females yielded similar figures in the Apenheul group (Aureli *et al.*, 1997), and the same was found in the semi-free ranging Kintzheim population (Figure 3).

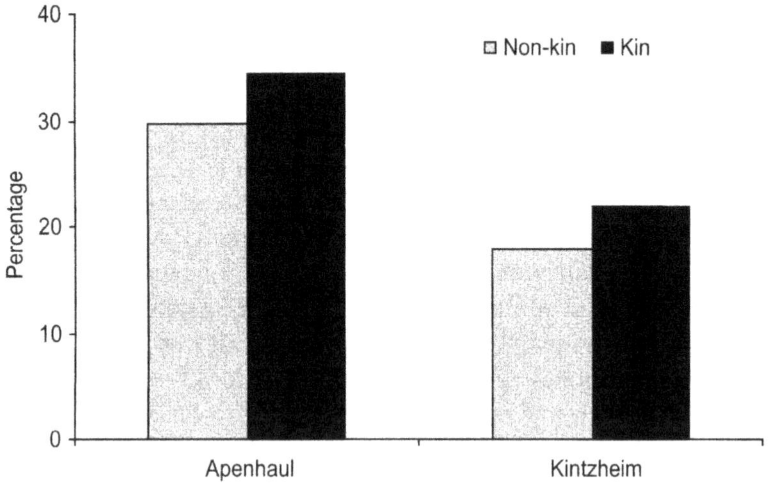

Figure 3. Conciliatory tendencies for kin and non-kin females. Conciliatory tendencies are calculated according to Veenema *et al.* (1994) from dyadic conflicts occurred between females at least 3.5-yr old (individual data pooled, N = 107 PC-MC pairs in Kintzheim and 158 in Apenheul).

The comparison of the different social styles observed in macaque species indicates that they represent covariant sets of characters (Thierry, 2004). In other words, the organisation of each species ranges within a sociospace defined by the interconnections of social characters. Barbary macaques fit well within this scheme. They display elaborate means of managing conflicts of interest, appeasing each other and decreasing social tension. Kinship bias and dominance asymmetry between conspecifics are moderate. Individuals may establish good relationships with most group members, and this applies even to relationships between adult males or between females from different matrilines. These patterns produce the tolerant style typical of Barbary macaques, allowing a strong development of alloparental behaviour, which in turn extends the socialisation network of infants and shape this way the relationships of the next generation. At present, however, we cannot specify the evolutionary determinants having driven Barbary macaques to their present state of social relationships. Several attempts have been made to correlate the contrasting social styles of macaques with the main ecological features of their habitats, predation risk and food distribution. These socioecological models still fail to account for the interspecific variations observed in macaque organisation (Ménard, 2004; Thierry, 2004).

Conditions of mating competition

Macaques may be classified into two main categories according to their mating relationships (Thierry, 2006). In tropical species reproduction occurs year round and since there is usually no more than one sexually receptive female at any given time, males and females form long-lasting consortships in which one male typically follows and mates with the female during days or weeks, excluding other males from reproduction. The close guarding exerted by the male leaves little room for female mate choice. This contrasts with species living in temperate regions in which most reproductive females cycle and become receptive within a two- to three-month autumn period. In such seasonal breeders no male is able to monopolise all fertile females and the duration of associations between males and females is short, typically between some hours and a few days. Males may shift from one female to another or supplant a lower-ranking male from the proximity of a female. Conversely a female may exert mate choice, refusing to copulate with a male and accepting another within minutes, even a lower-ranking one (Small, 1989; Huffman, 1991; Bercovitch, 1997; Soltis *et al.*, 1997). Thus, reproductive seasonality generally shapes mating patterns by influencing the operational sex ratio, *i.e.* ratio of adult males to fertile females.

Barbary macaques do not really fit this pattern. They breed seasonally and females do not associate with males for a long time, but the duration of associations between sexes is definitely briefer than in other species: most bouts last between one and twenty minutes and they rarely extend beyond one hour (Taub, 1980b; Small, 1990c; Kuester and Paul, 1992). Receptive females are almost exclusively responsible for the exchange of sexual partners (Figure 4). A male induces a female to join him by approaching and displaying courtship signals, or by following a couple and waiting for the female to end the association. Females often break off soon after copulation; indeed they determine the termination of most sexual associations. They mate with multiple males in rapid succession, commonly copulating with a majority of available group males in a day. No consistent rules appear in the way they select males, as they appear to indiscriminately copulate with males regardless their age, dominance rank and friendship (Taub, 1980b; Small, 1990c; Kuester and Paul, 1992). Whereas fully adult males generally avoid aggressively opposing each other, they do not hesitate to chase subadult males. Sneaking is the primary mating tactic of the latter; they usually go out of the view of adult males to mount females (Kuester and Paul, 1989, 1992).

The correlation between dominance rank and reproductive success is weak in adult males, as it is true for other seasonally breeding species (Paul, 2004). Since there is no link between paternity and the infant caretaking activity of males, it is unlikely that the reproductive behaviour of females is a strategy to protect their infants. They do not establish special relationships with the fathers of their offspring and confusing paternity cannot be very effective since each male cares for only a few infants (*cf.* above). The explanation of these patterns can be found in the context of mating competition (Small, 1990c). The operational sex ratio primarily determines mating patterns, hence the promiscuous reproductive style of Barbary macaques. The species-specific social style additionally influences the management of competition. In species like rhesus and Japanese macaques, strong dominance asymmetries allow dominant adult males to easily disrupt the consort couples of lower-ranking ones. Their mating patterns consequently represent a mix between contest and scramble. On the contrary, the balanced dominance relationships of male Barbary macaques typically prevent them from challenging their rivals by performing overt aggression. This stalemate process freezes the competition between adult males and allows oestrous females to behave unrestrained, resulting in the scrambling nature of the mating system of Barbary macaques (Small, 1990c; Kuester and Paul, 1992; Preuschoft *et al.*, 1998).

The demographic structure of Barbary macaques is characterised by an almost even adult sex ratio. This is linked to a late dispersal of males compared to other macaque species: two thirds of them disperse after seven years of age or remain in their natal groups (Ménard and Vallet, 1996; Kuester and Paul, 1999). Such features are generally attributed to the high levels of social tolerance observed in the species, which result from the conjunction between regular appeasement behaviours, relaxed dominance relationships and scramble mating patterns. Heightened interindividual tolerance should not be taken to mean that competition is low, however. Males may inflict significant wounds on each other during the mating season, and it is likely that mate competition levels in Barbary macaques are similar to those of other macaques (Kuester and Paul, 1992). Thus, the negative consequences of competition are managed by a suite of specific behaviour patterns, such as peaceful interventions in conflicts and triadic male-infant interactions, which allow large numbers of males to successfully deal with the intense conflicts of interest induced by the sexual context, but at the cost of being unable to control the reproductive behaviour of females.

Figure 4. Mating behaviours in wild Barbary macaques (Morocco, Middle Atlas). Females may roam freely (a) and invite males to mate (b) in the presence of others (c). Photographs by B. Thierry.

Conclusion

Morphological and molecular evidence has established the Barbary macaques as the species closest to the root of the phylogenetic tree of the genus *Macaca* (Fooden, 1976; Delson, 1980; Hoelzer and Melnick, 1996). Consistent with their geographical location in North Africa, it is the extant species most similar to the ancestor of macaques. Variations in the social styles of macaques have been shown to correlate with their phylogeny (Thierry *et al.*, 2000) and by tracing each of the behavioural characters on the phylogenetic tree, it is possible to recognise their most ancient states and reconstruct the typical ancestral organisation of macaques. The resulting set of characters, which may be considered as the ancestral state, closely matches grade 3 on the 4-grade scale, thus corresponding closely to the organisation of Barbary macaques (grade 3). In contrast, rhesus and Japanese macaques, which are among the taxa that have appeared most recently, and can be considered as the most derived species, belong to grade 1 on the social scale. Their intolerant style of relations is often qualified 'despotic' and they indeed display quite a 'barbarian' way of life in comparison to Barbary macaques (de Waal and Luttrell, 1989; Aureli *et al.*, 1997). In other words, that means that rhesus and Japanese macaques no longer use many of the affiliative behaviours that allow Barbary macaques to cope with social competition, but instead rely on open contests, strong hierarchies and kin support to rule their social life. At present, we are reaching a fair understanding of how social styles move into the macaque sociospace, but we still do not know which selective processes and ecological factors are liable to move them in one direction or another (Ménard, 2004; Thierry, 2004). We know that Barbary macaques are not 'barbarian', but future studies need to reveal the advantages and disadvantages of their tolerant style.

References

Aureli, F. and de Waal, F.B.M. (eds.), 2000. *Natural conflict resolution.* University of California Press, Berkeley.

Aureli, F. and van Schaik, C.P. 1991a. Post-conflict behaviour in long-tailed macaques (*Macaca fascicularis*): I. The social events. *Ethology* **89:** 89-100.

Aureli, F. and van Schaik, C.P. 1991b. Post-conflict behaviour in long-tailed macaques (*Macaca fascicularis*): II. Coping with the uncertainty. *Ethology* **89:** 101-114.

Aureli, F., Cozzolino, R., Cordischi, C. and Scucchi, S. 1992. Kin-oriented redirection among Japanese macaques: an expression of a revenge system? *Animal Behaviour* **44:** 283-291.

Aureli, F., Das, M. and Veenema, H.C., 1997. Differential kinship effect on reconciliation in three species of macaques (*Macaca fascicularis, M. fuscata*, and *M. sylvanus*). *Journal of Comparative Psychology* **111:** 91-99.

Aureli, F., Das, M., Verleur, D. and van Hooff, 1994. Postconflict social interactions among Barbary macaques (*Macaca sylvanus*). *International Journal of Primatology* **15:** 471-485.

Bercovitch, F.B., 1997. Reproductive strategies of rhesus macaques. *Primates* **38:** 247-263.

Chapais, B., 2004. How kinship generates dominance structures: a comparative perspective. In *Macaque societies: A model for the study of social organization*, B. Thierry, M. Singh and W. Kaumanns (eds.), Cambridge University Press, Cambridge, pp. 186-204.

Deag, J.M. and Crook, J.H., 1971. Social behaviour and 'agonistic buffering' in the wild Barbary macaque *Macaca sylvana*. L. *Folia Primatologica* **15:** 183-200.

Deag, J.M., 1977. Aggression and submission in monkey societies. *Animal Behaviour* **25:** 465-474.

Deag, J.M., 1980. Interactions between males and unweaned Barbary macaques: testing the agonistic buffering hypothesis. *Behaviour* **75:** 54-81.

Delson, E., 1980. Fossil macaques, phyletic relationships and a scenario of deployment. In *The macaques: Studies in ecology, behavior and evolution*, D.G. Lindburg (ed.), van Nostrand Reinhold, New York, pp. 10-30.

Deng, Z.Y., 1993. Social development of infants of *Macaca thibetana* at Mount Emei, China. *Folia Primatologica* **60:** 28-35.

de Waal, F.B.M. and Luttrell, L.M., 1989. Toward a comparative socioecology of the genus *Macaca*: different dominance styles in rhesus and stumptail macaques. *American Journal of Primatology* **19:** 83-109.

de Waal, F.B.M. and van Roosmalen, A., 1979. Reconciliation and consolation in chimpanzees. *Behavioral Ecology and Sociobiology* **5:** 55-66.

Fooden, J., 1976. Provisional classification and key to the living species of macaques (Primates: *Macaca*). *Folia Primatologica* **25:** 225-236.

Hoelzer, G.A. and Melnick, D.J., 1996. Evolutionary relationships of the macaques. In *Evolution and ecology of macaque societies*, J.E. Fa and D.G. Lindburg (eds.), Cambridge University Press, Cambridge, pp. 3-19.

Huffman, M.A., 1991. Mate selection and partner preferences in female Japanese macaques. In *The monkeys of Arashiyama: Thirty-five years of research in Japan and the West*, L.M. Fedigan and P.J. Asquith (eds.),

State University of New York Press, Albany, NY, pp. 101-122.

Judge, P.G., 1982. Redirection of aggression based on kinship in a captive group of pigtail macaques (abstract). *International Journal of Primatology* **3:** 301.

Kazem, A.J.N. and Aureli, F. 2005. Redirection of aggression: multiparty signalling within a network? In *Animal communication networks*, P.K. McGregor (ed.), Cambridge University Press, Cambridge, pp. 191-218

Kuester, J. and Paul, A., 1989. Reproductive strategies of subadult Barbary macaque males at Affenberg Salem. In *Sociobiology of reproductive strategies in animals and humans*, E.O.E. Rasa, C. Vogel and E. Voland (eds.), Chapman and Hall, London, pp. 93-109.

Kuester, J. and Paul, A., 1992. Influence of male competition and female mate choice on male mating success in Barbary macaques (*Macaca sylvanus*). *Behaviour* **120:** 192-217.

Kuester, J. and Paul, A., 1999. Male migration in Barbary macaques (*Macaca sylvanus*) at Affenberg Salem. *International Journal of Primatology* **20:** 85-106.

Lahiri, R.K. and Southwick, C.H., 1966. Parental care in *Macaca sylvana. Folia Primatologica* **4:** 257-264.

Maestripieri, D., 1994. Social structure, infant handling, and mothering styles in group-living Old World monkeys. *International Journal of Primatology* **15:** 531-553.

Marengo, C., 2002. *Les facteurs d'intervention dans les conflits chez le macaque berbère* (*Macaca sylvanus*). Diplôme d'Etudes Approfondies, Université Louis Pasteur, Strasbourg.

Ménard, N. and Vallet, D., 1996. Demography and ecology of Barbary macaques (*Macaca sylvanus*) in two different habitats. In J.E. Fa and D.G. Lindburg (eds.), *Evolution and Ecology of Macaque Societies*, Cambridge University Press, Cambridge, pp. 106-131.

Ménard, N., 2004. Do ecological factors explain variation in social organization? In *Macaque societies: A model for the study of social organization*, Thierry, B., Singh, M. and Kaumanns, W. (eds.), Cambridge University Press, Cambridge, pp. 237-262.

Ménard, N., von Segesser, F., Scheffrahn, W., Pastorini, J., Vallet, D., Gaci, B., Martin, R.D. and Gautier-Hion, A., 2001. Is male-infant caretaking related to paternity and/or mating activities in wild Barbary macaques (*Macaca sylvanus*)? *Comptes Rendus de l'Académie des Sciences de Paris III* **324:** 601-610.

Ogawa, H., 1995. Bridging behavior and other affiliative interactions among male Tibetan macaques (*Macaca thibetana*). *International Journal Primatology* **16:** 707-727.

Paul, A., 1989. Determinants of male mating success in a large group of

Barbary macaques (*Macaca sylvanus*) at Affenberg Salem. *Primates* **30:** 461-476.

Paul, A., 2004. Dominance and paternity. In *Macaque societies: a model for the study of social organization*, B. Thierry, M. Singh and W. Kaumanns (eds.), Cambridge University Press, Cambridge, pp. 131-134.

Paul, A. and Kuester, J. 1987. Dominance, kinship and reproductive value in female Barbary macaques (*Macaca sylvanus*). *Behavioral Ecology and Sociobiology* **21:** 323-331.

Paul, A. and Kuester, J., 1996. Infant handling by female Barbary macaques (*Macaca sylvanus*) at Affenberg Salem: testing functional and evolutionary hypotheses. *Behavioral Ecology and Sociobiology* **39:** 133-145.

Paul, A., Kuester, J. and Arnemann, J., 1996. The sociobiology of male-infant interactions in Barbary macaques, *Macaca sylvanus*. *Animal Behaviour* **51:** 155-170.

Petit O. and Thierry B., 1994. Aggressive and peaceful interventions in conflicts in Tonkean macaques. *Animal Behaviour* **48:** 1427-1436.

Petit, O. and Thierry, B., 2000. Do impartial interventions occur in monkeys and apes? In *Natural Conflict Resolution*, Aureli, F. and de Waal, F.B.M. (eds.), University of California Press, Berkeley, pp. 267-269.

Preuschoft, S., 1992. 'Laughter' and 'smile' in Barbary macaques (*Macaca sylvanus*). *Ethology* **91:** 220-236.

Preuschoft, S., Paul, A. and Kuester, J., 1998. Dominance styles of female and male Barbary macaques (*Macaca sylvanus*). *Behaviour* **135:** 731-755.

Preuschoft, S. and van Schaik, C.P., 2000. Dominance and communication. In *Natural conflict resolution*, F. Aureli and F.B.M. de Waal (eds.), University of California Press, Berkeley, pp. 77-105.

Prud'homme, J. and Chapais, B., 1993. Rank relations among sisters in semi-free ranging Barbary macaques (*Macaca sylvanus*). *International Journal of Primatology* **14:** 405-420.

Schino, G., Aureli, F., Ventura, R. and Troisi, A., 2004. A test of the cross-generational transmission of grooming preferences in macaques. *Ethology* **110:** 137-146.

Scucchi, S., Cordischi, C., Aureli, F. and Cozzolino, R. 1988. The use of redirection in a captive group of Japanese monkeys. *Primates* **29:** 229-236.

Small, M.F., 1989. Female choice in nonhuman primates. *Yearbook of physical Anthropology* **32:** 103-127.

Small, M.F., 1990a. Alloparental behaviour in Barbary macaques, *Macaca sylvanus*. *Animal Behaviour* **39:** 297-306.

Small, M.F., 1990b. Social climber: independent rise in rank by a female

Barbary macaque (*Macaca sylvanus*). *Folia Primatologica* **55:** 85-91.
Small, M.F., 1990c. Promiscuity in Barbary macaques (*Macaca sylvanus*). *American Journal of Primatology* **20:** 267-282.
Smith, E.O. and Peffer-Smith, P.G., 1982. Triadic interactions in captive Barbary macaques (*Macaca sylvanus*, Linnaeus, 1758): "agonistic buffering"? *American Journal of Primatology* **2:** 99-107.
Soltis, J., Mitsunaga, F., Shimizu, K., Yanagihara, Y. and Nozaki, M., 1997. Sexual selection in Japanese macaques I: female mate choice or male sexual coercion. *Animal Behaviour* **54:** 725-736
Taub, D.M., 1980a. Testing the 'agonistic buffering' hypothesis. I. The dynamics of participation in the triadic interaction. *Behavioral Ecology and Sociobiology* **6:** 187-197.
Taub, D.M., 1980b. Female choice and mating strategies among wild Barbary macaques (*Macaca sylvanus* L.). In *The macaques: Studies in ecology, behavior, and evolution*, D.G. Lindburg (ed.), van Nostrand Reinhold, New York, pp. 287-344.
Taub, D.M., 1984. Male caretaking among wild Barbary macaques (*Macaca sylvanus*). In *Primate paternalism*, D.M. Taub (ed.), van Nostrand Reinhold, New York, pp. 20-55.
Thierry, B., 1990. L'état d'équilibre entre comportements agonistiques chez un groupe de macaques japonais (*Macaca fuscata*). *Comptes-rendus de l'Académie des Sciences de Paris* 310 III: 35-40.
Thierry, B., 2000. Covariation of conflict management patterns across macaque species. In *Natural conflict resolution*, F. Aureli and F.B.M. de Waal (eds.), University of California Press, Berkeley, pp. 106-128.
Thierry, B., 2004. Social epigenesis. In *Macaque societies: A model for the study of social organization*, B. Thierry, M. Singh and W. Kaumanns (eds.), Cambridge University Press, Cambridge, pp. 267-290.
Thierry, B., Iwaniuk, A.N. and Pellis, S.M., 2000. The influence of phylogeny on the social behaviour of macaques (Primates: *Cercopithecidae*, genus *Macaca*). *Ethology* **106:** 713-728.
Thierry, B., Singh, M. and Kaumanns, W. (eds.), 2004. *Macaque societies: A model for the study of social organization,* Cambridge University Press, Cambridge.
Thierry, B., 2006. The macaques: a double-layered social organization. In *Primates in perspective*, C. Campbell, A. Fuentes, K.C. MacKinnon, M. Panger, and S. Bearder (eds.), Oxford University Press, Oxford, pp. 224-239.
Timme, A., 1995. Sex differences in infant integration in a semifree-ranging group of Barbary macaques (*Macaca sylvanus*, L. 1758) at Salem, Germany. *American Journal of Primatology* **37:** 221-231.
Veenema, H.C., Das, M. and Aureli, F., 1994. Methodological corrections

for the study of reconciliation. *Behavioural Processes* **31:** 29–38.
Widdig, A., 2000. Coalition formation among male Barbary macaques (*Macaca sylvanus*). *American Journal of Primatology* **50:** 37-51.
Zhao, Q.K., 1996. Male-infant-male interactions in Tibetan macaques. *Primates* **37:** 135-143.

Kinship and behaviour in Barbary macaques

A Paul

Abt. Soziobiologie/Anthropologie, Institut für Zoologie, Anthropologie und Entwicklungsbiologie, Universität Göttingen, Berliner Straße 28, 37073 Göttingen, Germany

Introduction

Ever since Japanese primatologists started their pioneering long-term research on the social life of Japanese macaques shortly after World War II (Kawai 1958, Kawamura 1958), macaques have played a prominent role in research on kinship and behaviour. Macaques were among the first species used to test Hamilton's (1964) theory of inclusive fitness (Kaplan 1977; Kurland 1977; Massey 1977), and they were also among the first species suggested to be able to recognise kin in the absence of familiarity (Wu *et al.* 1980). Whether the latter is true remains a matter of debate (see below), but more than 40 years of research have clearly demonstrated that (maternal) kinship is one of the most important determinants structuring the social life of macaques and other primates. Kinship influences a wide variety of behavioural decisions, including group membership and dispersal patterns, dominance relationships, spatial proximity, grooming interactions, co-feeding, conflict intervention, alliance formation, reconciliation, mate choice, and allomaternal care (reviewed in Bernstein 1988; Chapais and Berman 2004a; Gouzoules 1984; Kapsalis 2004; Walters 1987; Silk 2002). Moreover, recent evidence also indicates that kinship networks have adaptive value, since long-term research on wild female baboons has shown that social bonds enhance infant survival (Silk *et al.* 2003).

Nevertheless, as noted by Chapais and Berman (2004a), kinship also remains, to some extent, a black box. Mechanisms of kin recognition as well as its possible limits, are still not well understood, and the same appears to be true for the causes and consequences of interspecies differences in kin orientation. Current evidence, although limited, indicates that kinship plays a more fundamental role in some species than in others (Chapais and Berman 2004b), including even closely related species such as the macaques (Thierry 2004). While all macaques live in superficially similar social systems, they vary profoundly in their dominance style (*sensu* de Waal 1989), conflict management (*sensu* Aureli and de Waal 2000), and other behavioural traits (Thierry 2000). Based on these differences, Thierry (2000) proposed a four-grade scale classification of the macaques, where the most despotic or intolerant species (rhesus and Japanese macaques) represent grade 1, and the most egalitarian or tolerant species (Sulawesi macaques) represent grade 4. Thierry acknowledged that it is difficult to group intermediate species, but comparative data on conciliatory tendencies from nine species of macaques prompted him to place Barbary macaques near the egalitarian end of the continuum, in grade 3 (see Thierry 2006; this volume, for additional arguments in support of this view).

To re-consider the position of Barbary macaques in this scheme, a task complicated by the fact that males and females of this species exhibit different dominance styles (Preuschoft *et al.* 1998), is beyond the scope of this chapter (see Thierry 2004, for a discussion, and Sterck *et al.* 1997, for a slightly different categorization of primate social relationships). Nevertheless, Thierry's classification has implications that are important here since egalitarian species appear to be generally much less kin oriented than despotic species (Thierry 1990, 2000, 2004; see also Chapais and Berman 2004b, Butovskaya 1993, 2004). Consequently, it has been assumed that Barbary macaques show little or no kin bias in affiliative interactions (Kapsalis 2004). While this may be true for some types of interactions (Ménard *et al.* 2001; Paul *et al.* 1992a), it certainly does not characterise the Barbary macaque as a species that is not kin oriented, however. In fact, kinship affects the behaviour of Barbary macaques in a variety of contexts, and the aim of this chapter is to review what is known about the impact of kinship on the social life of Barbary macaques.

Kinship and social organization

Like many other cercopithecines, Barbary macaques live in multi-male, multi-female groups, ranging in size from 13 to 88 animals in the wild (Ménard 2002), and up to 200 individuals in food-enhanced settings (Paul and Kuester 1988). Moreover, as in other cercopithecines, kinship exerts a strong influence on group membership: females typically remain in their natal group for their entire life, while many males emigrate around the time of puberty and breed in other social groups (Paul and Kuester 1985, 1988; Kuester and Paul 1999). Rates of male migration are, however, lower than in other macaques: during a 20-year study on male migration in the Salem Barbary macaque population, only one third of all males eventually left their natal group (Kuester and Paul 1999). Similar data exist for wild groups, in which about 25 to 60 percent of all males stay in their natal group for life (Ménard and Vallet 1993a, 1996).

Yet, as revealed by the Salem study, the decision whether to leave or not was affected by matrilineal kin networks. Emigrants were found to have more female relatives, especially sisters, in their natal group than non-dispersers. Males without female relatives in their natal group almost never emigrated. Co-residence with older brothers yielded the opposite effect: males with older brothers in their natal group tended to stay, whereas males without older brothers left. Moreover, maternal relatives were significantly overrepresented among co-migrating males, suggesting that a male's decision to leave or not also affects the respective decision of his male relatives. Paternal brothers, on the other hand, did not co-migrate more often than expected by chance (Kuester and Paul 1999).

Kinship is an important determinant of male behaviour: a two-year old male huddles with his one-year old maternal brother. Photograph by A. Paul.

Analyses of group fissions corroborate the general picture and add some interesting details. Fissions by which one group splits to form two or more smaller daughter groups have been observed in a number of macaques (Okamoto 2004), including Barbary macaques (Kuester and Paul 1997; Ménard and Vallet 1993b; Prud'homme 1991). As in other cercopithecines, fissions tend to separate female kin groups, thereby increasing the genetic relatedness between the females of the new groups. During several group fissions in the Salem Barbary macaque population (Kuester and Paul 1997), immature females always joined their mothers, and mature females from the same matriline almost always stayed together – even if these matrilines were rather large (up to 15 females aged four years and older) and mean coefficients of relatedness rather low. Matriline splitting occurred in only two out of 43 matrilines with at least two mature females (4.8%). Matriline splitting has also been observed among female rhesus macaques on Cayo Santiago (Chepko-Sade and Sade 1979), and here the proportion was substantially larger (12.9% of all matrilines) – which is at variance with the view that rhesus macaque females are more kin oriented that Barbary macaque females.

Males from the same matriline also tended to stay together during group fissions, but here the proportion of within-matriline splitting was higher. Out of 35 matrilines with at least two males aged two years and older, within-matriline splitting occurred in nine cases (26%), and these matrilines had significantly more male members than those in which all males remained together (Kuester and Paul 1997).

Decisions of natal males as to which group to join during a fission also depended on the behaviour of their female kin. Many males, whose female kin collectively emigrated, remained in their natal group; to distinguish these males from true natal males who did not become separated from their female kin after fission, Kuester and Paul termed these males "semi-natal" males. The main difference between semi-natal and natal males was the following: semi-natal males had significantly more sisters and more female relatives overall, than natal males, and their families were female biased, that is, they had more sisters than brothers. Families of natal males, on the other hand, were male-biased: they had more brothers than sisters.

In contrast to maternal kinship, paternal kinship did not appear to influence individual decisions during group fissions. Separation of members from the same patriline was common, not only among larger patrilines (with at least four patrilineally related individuals), where within-patriline splitting almost always (94%) occurred, but also among smaller patrilines (52%).

Dominance acquisition, agonistic support and reconciliation

Rules of rank "inheritance" have once been regarded as "the heart of macaque social structure" (Schulman and Chapais 1980), but given the observed interspecies differences in dominance style (Thierry 2000), it is not surprising to see that these rules are more variable than originally suggested. According to the "classical" cercopithecine model, a female's rank is entirely predictable from knowledge of her kinship bonds (Chapais 2004): A daughter eventually outranks all females that rank below her mother and remains subordinate only to her mother. Hence, young females not only outrank older unrelated females subordinate to their mothers, they also outrank their older sisters, a phenomenon known as "youngest ascendancy" or "Kawamura principle" (after Kawamura 1958). Rank relations between sisters again determine the relative rank relations of their respective daughters, rank relations between aunts and nieces, cousins, and so on, giving rise to highly stable nepotistic hierarchies or matrilineal dominance structures. Rank relations among males are also affected by their mothers' ranks, but, partly because males habitually leave their natal group, less so than among females (see, e.g. Berard 1989, for rhesus macaques).

Barbary macaques do not fully conform to this classical model. Studies by Paul and Kuester (1987) and Prud'homme and Chapais (1993a, b) revealed that the dominance order of female Barbary macaques is clearly matrilineally structured, but, as noted by Chapais (2004), the level of nepotism appears to be lower than in the classical type. There are two reasons for this suggestion. First, at least in large groups, the process of rank acquisition is slower than, for example, in rhesus or Japanese macaques. While in these species even juvenile females commonly outrank adult females from lower ranking matrilines (Datta 1983; Chapais *et al.* 2001), most Barbary macaque females are able to do this only until they are themselves close to them in size. The fact that it may take up to eight years until a female reaches her eventual rank position (Paul and Kuester 1987) clearly indicates that there is an individualistic element in an otherwise nepotistic hierarchy. Old, post-reproductive matriarchs that were outranked not only by their daughters, but also by younger, unrelated females also underline this argument. Second, youngest ascendancy appears to be the exception rather than the rule among Barbary macaques, suggesting that the mother's involvement in rank relations between her daughters is not as strong as the classical model predicts. This does not mean that young Barbary macaque females do not receive support from their kin during conflicts with older sisters or unrelated females. Indeed, detailed observations by Prud'homme revealed that kin intervened disproportionately more often than non-kin in conflicts between sisters, and that this support was always in favour of the younger sister. However, only very young females (mostly yearlings) received support from their kin (Prud'homme and Chapais 1993a).

The main factor accounting for the failure of young female Barbary macaques to outrank their older sisters is, according to Prud'homme and Chapais (1993a), not the lack of kin support, but the lack of non-kin support. Support from mothers and other close relatives on behalf of their kin, they argue, does occur, but is rarely frequent and intense enough to foster rebellion of young females against their older sisters. As a consequence, these young females usually do not prompt non-kin females to join them opportunistically against their older sisters (see Chapais 2004, for a more detailed treatment of this argument). In any case, deviations from the Kawamura principle are not restricted to Barbary macaques. Youngest ascendancy appears to be irrelevant among Tonkean macaques (Thierry 2004), and deviations from the rule have also been reported from more nepotistic (and despotic) species such as Japanese macaques (e.g. Hill and Okayasu 1995, Koyama 2003), or Tibetan macaques (Berman *et al.* 2004).

Kin support in Barbary macaques is not restricted to very young females, however. Research in the Salem colony revealed that adult females from large matrilines were most often supported by other females from their own matriline, while females from small matrilines received most support from unrelated females (Steuckardt 1991).

Low rates of kin support in small matrilines have also been reported by Aureli *et al.* (1997), who found that among the Barbary macaques at Apenheul (Netherlands) support on behalf of close kin (mother-offspring and maternal siblings) accounted for only 7% of all observed interventions, whereas among similar groups of long-tailed and Japanese macaques the respective values were between 40 and 50%.

Matrilineal kinship also influences rank acquisition of male Barbary macaques, although to a lesser extent than it is the case in females. Kuester and Paul (1988) found in 8 out of 12 cohorts a significant positive correlation between the males' rank among their (juvenile or subadult) peers and their mothers' rank. Support from older brothers – if present — also appears to facilitate male rank acquisition (Kuester and Paul 1988), and anecdotal evidence indicates that coalitionary support may enable clusters of closely related adult males to achieve top rank positions in their natal group (W. Angst, personal communication, and unpublished observation). A systematic, but short-term study by Widdig *et al.* (2000) in one of the Salem groups also revealed that during conflicts, natal males preferentially support close kin. Males in this study intervened nearly twice as often in conflicts in which a maternal relative was involved than in conflicts involving only non-kin, and they intervened on behalf of close kin (brothers) more often than on behalf of more distant kin (uncles, nephews, cousins). However, as has been found in other studies, coalition formation was not restricted to kin, and most third party interventions during agonistic conflicts appeared to be governed by self-interests (Widdig *et al.* 2000; J. Kuester, unpublished data; see also Prud'homme and Chapais 1993b, for female Barbary macaques).

Possible influences of both maternal and paternal kinship on male-male relations have been investigated, to my knowledge, in only one study (an unpublished diploma thesis by Barbaranelli 1994). This study revealed that some, but not all, maternal brothers spent much time near one another, resulting in a non-significant difference between matrineally related and unrelated pairs of males, while patrilineally related males spent as much time together as unrelated males. Unfortunately, small sample sizes precluded any meaningful analyses of support patterns and other friendly interactions.

Comparative data on post-conflict social interactions show that, in contrast to Japanese and long-tailed macaques, post-conflict reunions among Barbary macaques do not take place more often after conflicts between related opponents (Aureli *et al.* 1997). Moreover, reconciliation between unrelated opponents occurred more than twice as often among Barbary macaques as among Japanese and long-tailed macaques, suggesting that Barbary macaques may rely more on non-kin as coalition partners than more despotic species. Nevertheless, data on affiliative interactions (allogrooming, huddling, play) revealed a similar kin bias in Barbary macaques as in the other two species (Aureli *et al.* 1997). Evidence for a post-conflict increase in affiliation between the victim and its kin ("consolation") is currently lacking for all macaque species (Aureli *et al.* 1994).

Allomaternal care

High rates of allomaternal care (at least for cercopithecines, cf. Ross and McLarnon 2000) and, especially, male care, are among the most prominent behavioural features of Barbary macaques. Infants spend, on average, about 20 percent of the daytime in physical contact with individuals other than their mothers (Paul 1999). While it is now well established that infant handling by males is not governed by kinship relations (Kuester and Paul 1986; Ménard *et al.* 2001; Paul 1984; Paul *et al.* 1992a, 1996), the situation is clearly different for infant handling by females (Paul and Kuester 1996). Although females handle unrelated infants quite often (about 40 percent of all episodes observed by Paul and Kuester 1996), there was a strong preference for (maternally) related infants, and involvement with infants correlated almost perfectly with the degree of relatedness (Figure 1).

In fact, even cousins and more distant relatives established preferential relationships with infants more often than expected by chance (values above 1 in Figure 1), while only unrelated females (and very old, post-reproductive great-grandmothers) fell below that level.

Sex and reproduction

A wealth of evidence indicates that primates generally avoid incest, and Barbary macaques are no exception to the rule (Kuester *et al.* 1994; Paul and Kuester 1985, 2004). Given that inbreeding, under most circumstances, tends

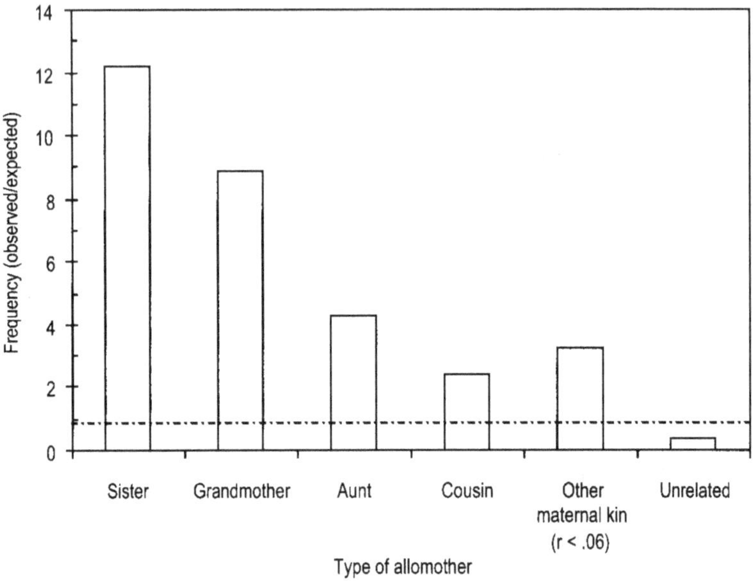

Figure 1. Preferential relationships between female allomothers and infants in the Salem Barbary macaque colony (source: Paul and Kuester 1996)

to be costly (e.g. Crnokrak and Roff 1999; Hamilton 1993), this may not be surprising, but analyses of Barbary macaque sexual interactions have revealed some interesting details. First, sexual interactions between maternally related individuals are extremely rare even though many males remain (and breed) in their natal group. This strongly suggests that inbreeding avoidance is not simply the result of sex-biased dispersal, but also of a psychological inclination of one or both sexes against sexual intercourse with close kin. Second, this psychological inclination appears to inhibit not only close maternal kin, but also more distantly related individuals. Sexual interactions between uncles and their nieces or aunts and their nephews and even between maternal cousins were still significantly less frequent than sexual interactions between unrelated individuals (Paul and Kuester 2004). Third, rare cases of unfamiliar maternal relatives revealed no mating inhibition, while sexual interactions between females and their former (unrelated) male caretakers were more or less absent. And finally, even close paternal kin such as fathers and daughters or paternal siblings did not avoid incestuous matings (Kuester *et al.* 1994). Nevertheless, at least one mechanism appears to prevent sibling incest: paternity analyses revealed that no female in the Salem colony ever produced an offspring with a same-aged natal peer, suggesting that Barbary macaques discriminate against natal members of their own age cohort. Moreover, age differences between natal fathers and mothers were significantly greater than between non-natal fathers and natal mothers (Paul and Kuester 1998).

Conclusions

Forty years after Hamilton's seminal paper (Hamilton 1964) questions about kin recognition, kin discrimination and the allocation of altruism still belong to the hottest topics in primate sociobiology (Chapais and Berman 2004a, b). Are primates able to recognise paternal kin in the absence of familiarity, for example? Several recently published studies suggest that the answer may be yes (Alberts 1999; Buchan *et al.* 2003; Smith *et al.* 2003; Widdig *et al.* 2001), but scepticism remains (see Paul and Kuester 2004; Rendall 2004). The data reviewed here clearly suggest that Barbary macaques are not able to recognise paternal kin. Instead, familiarity appears to be the main mechanism involved in kin recognition and kin discrimination. Yet, 'familiarity' is a vague concept: We are still far from a complete understanding of how this mechanism exactly works. Playback experiments have revealed that Barbary macaque mothers recognise their 4 to 8 week old offspring by acoustic signals alone (Hammerschmidt and Fischer 1998), and infants recognise their mothers' voice at the age of 10 weeks (Fischer 2004). Apart from that, little is known about mechanisms of kin recognition and kin discrimination. What is it that makes individuals 'familiar' with each other: quantity or quality of social interactions, timing, amount or duration of physical contact? Are there imprinting-like mechanisms, i.e. genetically canalised learning processes that shape partner preferences (or avoidance of certain mates) during certain critical periods? Research on humans offers promising perspectives (Bereczkei *et al.* 2004), but knowledge about nonhuman primates remains poor.

Are Barbary macaques less kin-oriented than other, less tolerant macaque species? Patterns of rank acquisition, agonistic support and reconciliation indeed suggest that Barbary macaques may be less nepotistic than, for example, Japanese macaques are. On the other hand, there is little evidence for the existence of a "generalised (non behaviour-specific) relatedness threshold" at r=0.25 for the preferential treatment of kin among Barbary macaques, as it has been suggested for Japanese macaques (Bélisle and Chapais 2001). Given that kin discrimination in Barbary macaques appears to go well beyond r=0.25, Barbary macaques seem even more kin-oriented than Japanese macaques are.

Finally, do kin-biases in behaviour enhance the inclusive fitness of the actors? Evidence is limited, although in the case of inbreeding avoidance the answer is likely to be yes. The same may be true for kin-biased support (cf. Chapais *et al.* 2001), or the decision to leave one's own maternal kin or to remain in the natal group (Kuester and Paul 1997, 1999). There is also evidence suggesting that kinship relations affect male reproductive success after natal dispersal, although the precise mechanism is unclear (Paul *et al.* 1992b). Whether kin biases in infant handling have been favoured by kin selection however, remains obscure (Paul and Kuester 1996).

Acknowledgments

Research in the Salem Barbary macaque colony was supported by the Deutsche Forschungsgemeinschaft (An 131/1-6, Pa 408/2-1, Vo 124/15-1). Many thanks to Walter Angst, Joachim Arnemann, Ellen Merz, Signe Preuschoft, Gilbert de Turckheim, and, last but not least, Jutta Kuester, without whose long-term collaboration and friendship this chapter would not have been possible.

References

Alberts, S. C. 1999. Paternal kin discrimination in wild baboons. *Proceedings of the Royal Society London* B, **266**, 1501-1506.

Aureli, F. and de Waal, F. B. M., eds., 2000. *Natural Conflict Resolution*. Berkeley: The University of California Press.

Aureli, F., Das, M., Verleur, D. and van Hooff, J. A. R. A. M. 1994. Postconflict social interactions among Barbary macaques (*Macaca sylvanus*). *International Journal of Primatology*, **15**, 471-485.

Aureli, F., Das, M. and Veenema, H. C. 1997. Differential kinship effect on reconciliation in three species of macaques (*Macaca fascicularis, M. fuscata*, and *M. sylvanus*). *Journal of Comparative Psychology*, **111**, 91-99.

Barbaranelli, G. 1994. *Sozialbeziehungen zwischen adulten Berberaffenmännchen (*Macaca sylvanus L. 1758*)*. Unpublished Diploma thesis, University of Göttingen, Göttingen.

Bélisle, P. and Chapais, B. 2001. Tolerated co-feeding in relation to degree of kinship in Japanese macaques. *Behaviour*, **138**, 487-509.

Berard, J. 1989. Life histories of male Cayo Santiago macaques. *Puerto Rico Health Science Journal*, **8**, 61-64.

Bereczkei, T., Gyuris, P. and Weisfeld, G. E. 2004. Sexual imprinting in human mate choice. *Proceedings of the Royal Society* Lond. B, **271**, 1129-1134.

Berman, C. M., Ionica, C. S. and Li, J. 2004: Dominance style among *Macaca thibetana* on Mt. Huangshan, China. *International Journal of Primatology*, **25**, 1283-1312.

Bernstein, I. S. 1988. Kinship and behavior in nonhuman primates. *Behavior Genetics*, **18**, 511-524.

Buchan, J. C., Alberts, S. C., Silk, J. B. and Altmann, J. 2003. True paternal care in a multi-male primate society. *Nature*, **425**, 179-181.

Butovskaya, M. 1993. Kinship and different dominance styles in groups of three species of the genus *Macaca* (*M. arctoides, M. mulatta, M. fascicularis*). *Folia Primatologica*, **60**, 210-224.

Butovskaya, M. 2004. Social space and degrees of freedom. In: B. Thierry, M. Singh and W. Kaumanns (eds.), *Macaque Societies. A Model for the Study of Social Organization*. Cambridge: Cambridge University Press, pp. 182-185

Chapais, B. 2004. How kinship generates dominance structures: a comparative perspective. In: B. Thierry, M. Singh and W. Kaumanns (eds.), *Macaque Societies. A Model for the Study of Social Organization*. Cambridge: Cambridge University Press, pp. 186-204.

Chapais, B. and Berman, C. M. 2004a: Introduction: the kinship black box. In: B. Chapais and C.M. Berman (eds.), *Kinship and Behavior in Primates*. Oxford: Oxford University Press, pp. 3-11.

Chapais, B. and Berman, C. M. 2004b: Variation in nepotistic regimes and kin recognition: a major area for future research. In: B. Chapais and C.M. Berman (eds.), *Kinship and Behavior in Primates*. Oxford: Oxford University Press, pp. 477-489.

Chapais, B., Savard, L. and Gauthier, C. 2001. Kin selection and the distribution of altruism in relation to degree of kinship in Japanese macaques (*Macaca fuscata*). *Behavioral Ecology and Sociobiology*, **49**, 493-502.

Chepko-Sade, B. B. and Sade, D. S. 1979: Patterns of group splitting within matrilieal kinship groups: a study of social group structure in *Macaca mulatta* (Cercopithecidae: Primates). *Behavioral Ecology and Sociobiology*, **5**, 67-86.

Crnokrak, P. and Roff, D. A. 1999. Inbreeding depression in the wild. *Heredity*, **83**, 260-270.

Datta, S. B. 1983: Relative power and the acquisition of rank. In: R.A. Hinde (ed.), *Primate Social Relationships. An Integrated Approach*. Oxford: Blackwell Scientific Publications, pp. 93-103.

de Waal, F. B. M. 1989. Dominance «style» and primate social organization. In: V. Standen and R. A. Foley (eds.), *Comparative Socioecology: The Behavioural Ecology of Humans and other Mammals*. Oxford: Blackwell Scientific Publications, pp. 243-263.

Fischer, J. 2004. Emergence of individual recognition in young macaques. *Animal Behaviour*, **67**, 655-661.

Gouzoules, S. 1984: Primate mating systems, kin association and cooperative behavior: evidence for kin recognition ? *Yearbook of Physical Anthropology*, **27**, 99-134.

Hamilton W. D. 1964. The genetical theory of social behavior, I, II. *Journal of Theoretical Biology*, **7**, 1-52.

Hamilton, W. D. 1993. Inbreeding in Egypt and in this book: a childish perspective. In: N. Thornhill (ed.), *The Natural History of Inbreeding and Outbreeding. Theoretical and Empirical Perspectives*. Chicago: The University of Chicago Press, pp. 429-450.

Hammerschmidt, K. and Fischer, J. 1998. Maternal discrimination of offspring vocalisations in Barbary macaques (*Macaca sylvanus*). *Primates*, **39**, 231-236.

Hill, D. A. and Okayasu, N. 1995. Absence of 'youngest ascendancy' in the dominance relations of sisters in wild Japanese macaques (*Macaca fuscata yakui*). *Behaviour*, **132**, 367-379.

Kaplan, J. 1977. Patterns of fight interference in free-ranging rhesus monkeys. *American Journal of Physical Anthropology*, **47**, 279-287.

Kapsalis, E. 2004. Matrilineal kinship and primate behavior. In: B. Chapais and C.M. Berman (eds.), *Kinship and Behavior in Primates*. Oxford: Oxford University Press, pp. 153-176.

Kawai, M. 1958. On the system of social ranks in a natural troop of Japanese monkeys. *Primates*, **1**, 111-130.

Kawamura, S. 1958. Matriarchal social ranks in the Minoo-B-troop: a study of the rank system of Japanese monkeys. *Primates*, **1**, 149-156.

Koyama, N. 2003. Matrilineal cohesion and social networks in *Macaca fuscata*. *International Journal of Primatology*, **24**, 797-811.

Kuester, J. and Paul, A. 1986. Male-infant relationships in semifree-ranging Barbary macaques (*Macaca sylvanus*) of Affenberg Salem/FRG: testing the «male care» hypothesis. *American Journal of Primatology*, **10**, 315-327.

Kuester, J. and Paul, A. 1988. Rank relations of juvenile and subadult natal males of Barbary macaques (*Macaca sylvanus*) at Affenberg Salem. *Folia Primatologica*, **51**, 33-44.

Kuester, J. and Paul, A. 1997. Group fission in Barbary macaques (*Macaca sylvanus*) at „Affenberg Salem". *International Journal of Primatology*, **18**, 941-966.

Kuester, J. and Paul, A. 1999. Male migration in Barbary macaques (*Macaca sylvanus*) at Affenberg Salem. *International Journal of Primatology*, **20**, 85-106.

Kuester, J., Paul, A. and Arnemann, J. 1994. Kinship, familiarity and mating avoidance in Barbary macaques (*Macaca sylvanus*). *Animal Behaviour*, **48**, 1183-1194.

Kurland J. A. 1977. *Kin selection in the Japanese monkey*. Basel: Karger.

Massey, A. 1977. Agonistic aids and kinship in a group of pigtail macaques. *Behavioral Ecology and Sociobiology*, **2**, 31-40.

Ménard, N. 2002. Ecological plasticity of Barbary macaques (*Macaca sylvanus*). *Evolutionary Anthropology*, Suppl **1**, 95-100.

Ménard, N. and Vallet, D. 1993a. Population dynamics of *Macaca sylvanus* in Algeria: an 8-year study. *American Journal of Primatology*, **30**, 101-118.

Ménard, N. and Vallet, D. 1993b. Dynamics of fission in a wild Barbary macaque group (*Macaca sylvanus*). *International Journal of Primatology*, **14**, 479-

500.

Ménard, N. and Vallet, D. 1996. Demography and ecology of Barbary macaques (*Macaca sylvanus*) in two different habitats. In: J. E. Fa and D. G. Lindburg (eds.), *Evolution and Ecology of Macaque Societies*. Cambridge: Cambridge University Press, pp. 106-131.

Ménard, N., von Segesser, F., Scheffrahn, W., Pastorini, J., Vallet, D., Gaci, B., Martin, R. D. and Gautier-Hion, A. 2001. Is male-infant caretaking related to paternity and/or mating activities in wild Barbary macaques (*Macaca sylvanus*)? *Comptes Rendus de L'Academie des Sciences* III, **324**, 601-610.

Okamoto, K. 2004: Patterns of group fission. In: B. Thierry, M. Singh and W. Kaumanns (eds.), *Macaque Societies. A Model for the Study of Social Organization.* Cambridge: Cambridge University Press, pp. 112-116.

Paul, A. 1984. *Zur Sozialstruktur und Sozialisation semi-freilebender Berberaffen* (*Macaca sylvanus* L. 1758). PhD dissertation, Univ. of Kiel, Kiel.

Paul, A. 1999. The socioecology of infant handling in primates: Is the current model convincing? *Primates,* **40**, 33-46.

Paul, A. and Kuester, J. 1985. Intergroup transfer and incest avoidance in semifree-ranging Barbary macaques (*Macaca sylvanus*) at Salem (FRG). *American Journal of Primatology,* **8**, 317-322.

Paul, A. and Kuester, J. 1987. Dominance, kinship and reproductive value in female Barbary macaques (*Macaca sylvanus*) at Affenberg Salem. *Behavioral Ecology and Sociobiology*, **21**, 323-331.

Paul, A. and Kuester J. 1988. Life-history patterns of Barbary macaques (*Macaca sylvanus*) at Affenberg Salem. In: J.E. Fa and C.H. Southwick (eds.), *Ecology and Behavior of Food-Enhanced Primate Groups*. New York: Alan R Liss, pp 199-228.

Paul, A. and Kuester, J. 1996. Infant handling by female Barbary macaques (*Macaca sylvanus*) at Affenberg Salem: testing functional and evolutionary hypotheses. *Behavioral Ecology and Sociobiology*, **39**, 133-145.

Paul, A. and Kuester, J. 1998. Mate choice in Barbary macaques: what do mothers and fathers have in common? *Folia Primatologica,* **69**, 213-214.

Paul, A. and Kuester, J. 2004. The impact of kinship on mating and reproduction. In: B. Chapais and C. M. Berman (eds.), *Kinship and Behavior in Primates.* Oxford: Oxford University Press, pp. 271-291.

Paul, A., Kuester, J. and Arnemann, J. 1992a. DNA fingerprinting reveals that infant care by male Barbary macaques (*Macaca sylvanus*) is not paternal investment. *Folia Primatologica,* **58**, 93-98.

Paul, A., Kuester, J. and Arnemann J. 1992b. Maternal rank affects reproductive success of male Barbary macaques (*Macaca sylvanus*): evidence from DNA fingerprinting. *Behavioral Ecology and Sociobiology,* **30**, 337-341.

Paul, A. and Kuester, J. and Arnemann, J. 1996. The sociobiology of male-infant interactions in Barbary macaques, *Macaca sylvanus. Animal Behaviour,* **51**, 155-170.

Preuschoft, S., Paul, A. and Kuester, J. 1998. Dominance styles of female and male Barbary macaques (*Macaca sylvanus*). *Behaviour,* **135**, 731-755.

Prud'homme, J. 1991. Group fission in a semifree-ranging population of Barbary macaques (*Macaca sylvanus*). *Primates*, **32**, 9-22.

Prud'homme, J. and Chapais, B. 1993a. Rank relations among sisters in semi-free-ranging Barbary macaques (*Macaca sylvanus*). *International Journal of Primatology*, **14**, 405-420.

Prud'homme, J. and Chapais, B. 1993b. Aggressive interventions and matrilineal dominance relations in semifree-ranging Barbary macaques (*Macaca sylvanus*). *Primates*, **34**, 271-283.

Rendall, D. 2004. 'Recognizing' kin: mechanisms, media, minds, modules, and muddles. In: B. Chapais and C. M. Berman (eds.), *Kinship and Behavior in Primates*. Oxford: Oxford University Press, pp. 295-316.

Ross, C. and MacLarnon, A. 2000. The evolution of non-maternal care in anthropoid primates: a test of hypotheses. *Folia Primatologica,* **71**, 93-113.

Schulman, S.R. and Chapais, B. 1980. Reproductive value and rank relations among macaque sisters. *The American Naturalist*, **115**, 580-593.

Silk, J.B. 2002. Kin selection in primate groups. *International Journal of Primatology*, **23**, 849-876.

Silk, J. B., Alberts, S. C. and Altmann, J. 2003. Social bonds of female baboons enhance infant survival. *Science*, **302**, 1231-1234.

Smith, K., Alberts, S. C. and Altmann, J. 2003. Wild female baboons bias their social behavior towards paternal half-sisters. *Proceedings of the Royal Society* Lond. B, **270**, 503-510.

Sterck, E. H. M., Watts, D. P. and van Schaik, C. P. 1997. The evolution of female social relationships in nonhuman primates. *Behavioral Ecology and Sociobiology*, **41**, 291-309.

Steuckardt, A. 1991: *Der Einfluß der Familiengröße auf das Beziehungsnetz adulter Berberaffenweibchen* (*Macaca sylvanus* L. 1758). Unpublished diploma thesis, University of Göttingen. Göttingen

Thierry, B. 1990. Feedback loop between kinship and dominance: the macaque model. *Journal of Theoretical Bio*logy, **145**, 511-521.

Thierry, B. 2000. Covariation of conflict management patterns across macaque species. In: F. Aureli and F. B. M. de Waal (eds.), *Natural Conflict Resolution*. Berkeley: University of California Press, pp. 106-128.

Thierry, B. 2004. Social epigenesis. In: B. Thierry, M. Singh and W. Kaumanns (eds.), *Macaque Societies. A Model for the Study of Social Organization.* Cambridge: Cambridge University Press, pp. 267-290.

Walters, R. 1987. Kin recognition in nonhuman primates. In: D. J. C. Fletcher and C. D. Michener (eds.), *Kin Recognition in Animals*. New York: Wiley, pp. 359-393.

Widdig, A., Streich, W.J. and Tembrock, G. 2000. Coalition formation among male Barbary macaques (*Macaca sylvanus*). *American Journal of Primatology*, **50**, 37-51.

Widdig, A., Nürnberg, P., Krawczak, M., Streich, W. J. and Bercovitch, F. B. 2001. Paternal relatedness and age proximity regulate social relationships among adult female rhesus macaques. *Proceedings of the National Academy of Sciences* USA, **98**, 13769-13773.

Wu, H. M. H., Holmes, W. G., Medina, S. R. and Sackett, G. P. 1980. Kin preference in infant *Macaca nemestrina*. *Nature*, **285**, 225-227.

Vocal communication in Barbary macaques: a comparative perspective

J Fischer and K Hammerschmidt
Research Group Cognitive Ethology, German Primate Centre, Kellnerweg 4, Göttingen 37077, Germany

Introduction

The cries of monkeys and apes had already inspired Charles Darwin to reason about the similarity and differences in the vocal expressions of nonhuman primates and humans (Darwin 1872). Understanding the evolution of speech is still a driving force in the study of nonhuman primate (hereafter: "primate") communication. Within this framework, a sound ethological research programme has emerged over recent decades that encompasses studies at an ontogenetic as well as a phylogenetic level, and in which investigations of the mechanisms underlying the production and perception of vocalizations have been complemented by studies of the function of calling behaviour. In other words, Tinbergen's four levels of analysis (Tinbergen 1963) have been applied to the vocal communication of monkeys and apes.

Much of the research on ontogenetic development, that is development between birth and death, has focused on the question whether the ability to produce certain vocalizations was a learned versus an innate process. While there is little evidence that nonhuman primates learn to produce their sounds through imitation, learning does seem to play an important role in the usage and comprehension of calls (for a review, see Fischer 2002).

In terms of the underlying mechanisms, scholars have investigated the neuronal substrate involved in the production (e.g., Jürgens 1979; Deacon 1992) and auditory processing of vocalizations (e.g. Rauschecker *et al.* 1995), while at the behavioural level, a prominent question has been whether monkey or ape calls refer to objects or events in the subjects' environment, and whether listeners categorise calls solely according to acoustic similarity or also according to the 'meaning' of calls (for reviews, see Seyfarth and Cheney 2003; Ghazanfar and Santos 2004).

A further set of studies has addressed whether calls provide information about the signaller's status or reproductive state. For instance, male Chacma baboons produce display calls. Both the frequency of calling and the structure of the calls are related to male rank and predict whether or not two males will escalate the contest into physical fighting (Kitchen *et al.* 2003; Fischer *et al.* 2004). Similarly, copulation calls of female yellow baboons and Barbary macaques vary with the reproductive state of the female, and also male mating partner (Semple 1998; Semple *et al.* 2002; Semple and McComb, this volume).

All of the studies mentioned above have looked at the variation of acoustic characteristics within a given species. Fewer studies have aimed at comparing the call characteristics taking into account the phylogenetic relationships of the species under consideration. An exception is the analysis of the Gibbon song, an elaborate series of calls performed either as a duet by a male and a female, or produced by only one of the sexes (Geissmann 2002). Since the gibbon song stands out from all other primate vocalizations, it lends itself to such targeted investigations. However, if one wishes to compare entire vocal repertoires of different species, things tend to become more complicated. The starting point of such analyses may either be the structure of the calls themselves, or their putative function, that is, in which contexts they tend to occur. For instance, one may compare the acoustic structure of 'alarm calls', 'contact calls', or 'food calls'. However, it may well be the case that a similarly structured call has taken on a different function in a different species. It is therefore also important to assess the structure of calls without making any a priori assumptions about their function. The advent of computer based analyses of sounds and the development of multivariate statistical tools allowed scholars to describe call structure in much detail, and also to establish the morphology of a vocal repertoire. However, there is no gold standard when it comes to the application of these techniques, and hence, most of the studies published have measured differing suites of variables which still renders comparisons across species difficult.

In the first part of this chapter, we will provide an overview of the vocal behaviour of Barbary macaques (*Macaca sylvanus* L.1758). We will first present spectrographic displays of the most common calls. In the second part of this chapter, we will review the results of an extended computer-based analysis of the morphology of the vocal repertoire. In the discussion, we aim to put our findings into an evolutionary perspective, comparing the call characteristics of Barbary macaques to those of other macaque species. For reasons outlined above, this comparison will necessarily be sketchy, but possibly motivate scientists to adopt a common framework for future such studies.

Methods

The studies reported here were carried out in an outdoor enclosure at Rocamadour, France (size: 15 ha). The enclosure is a visitor park where monkeys range freely while visitors are restricted to a path. Individuals are well habituated to human observers and are tattooed with an individual code on the inside of the thigh. The monkeys are provisioned with monkey chow provided in feeder huts, and with apples, grain, and seeds which are spread throughout the park. For details on park management and park size, see Turckheim and Merz (1984). We established five age-classes: age-class 1: infants up to the age of one year, 2. young juveniles of one to two years, 3. elder juveniles of two to three and a half years, 4. subadults: females of three and a half to five years and males of three and a half to seven years, and 5. adult females older than five years and adult males older than seven years of age

We conducted the recordings between 1987 and 2001 as part of a number of studies on the monkeys' vocal communication (overview in Todt *et al.* 1995). Calls presented in this paper were selected from a data base containing more than 15,000 calls from 92 individually identified subjects, 40 males and 52 females. Details of the analysis settings for the quantitative description of the repertoire are given in Hammerschmidt and Fischer 1998). Avisoft (version 3.93, 2001, R. Specht, Berlin) was used to generate the spectrograms shown in the present paper.

In order to analyse the structure of the repertoire, we conducted a cluster analysis (method: k-means cluster) on the standardised values (z-scores) of the acoustic variables. A cluster analysis is basically a method to sort objects on the basis of their similarity with regard to a whole number of features. With the method that we used, the process starts by treating the whole data set as one cluster. The analysis then seeks to partition this cluster into two clusters, and next into three, four, and so on. A priori, there is no knowledge of which case (or which call) belongs to which group or of how many clusters to expect. Hence, we calculated 29 cluster solutions (2-30 clusters) and assessed them separately according to a procedure described in Bacher (1994). In this procedure, the observed variation for each cluster solution is compared to the overall variation of the unpartitioned data set ("Zero-model"). In other words, we were looking for a cluster solution that yielded minimum variation within the clusters and maximum variation between the clusters. The total number of calls in this analysis was 8479.

We established 10 main social contexts with various specifications (for details, see (Hammerschmidt and Fischer 1998):

1. 'Mating': calls given in the mating season during copulation, attempted copulation, interrupted copulation, and after copulation.
2. 'Infant care': calls emitted by the care-giver during infant care, during both maternal and alloparental care.

3. 'Quiet situation': resting, foraging, locomotion.
4. 'Agonistic situations': various agonistic encounters including vocalizations from both the aggressor and recipient of aggression.
5. 'Disturbance': disturbance by the observer, by park staff, by a dog, by a snake.
6. 'Sleeping cluster formation': an infant or young juvenile attempts to achieve body contact with its mother or other subjects.
7. 'Infant in care': infant is with the mother or an alloparent.
8. 'Play': solitary play, rough-and-tumble play.
9. 'Contact': making or breaking contact with group of subjects huddled on the ground.
10. 'Infant search': calls recorded from females whose dead infant was taken away by the park staff, whose infant was with another individual in sight or out of sight of the mother. Females appeared to be looking for their infants or were trying to retrieve their infants from alloparental care.

Results

Qualitative description

Below, we provide spectrographic examples of calls given by Barbary macaques. The spectrograms document that typically, calls are given in series ('bouts') of several consecutive calls. These may either consist of a series of the same call type ('homotype series'), or of several different call types ('heterotype series'; Todt 1986).

Mating

Barbary macaque females are assertive and frequently approach males to initiate copulations. Once the male has mounted the female for copulation, the female often looks back to the male, grasps one of his legs, and/or utters a rhythmic series of low-frequency grunts, also termed 'mating call' (Todt *et al.* 1995). During peak receptivity, females sometimes utter a call that sounds similar to the mating call although the female is not engaged in any mating activity. These calls may be given when females attempt to initiate a copulation but the male does not respond, when a consort pair has been interrupted by a third party, and sometimes simply when the female is sitting by herself. An analysis revealed that there are structural differences in the single units according to whether the call was uttered during copulation or in another situation (Todt *et al.* 1995). Further analysis of the temporal and spectral characteristics of these calls also indicate that there are differences in relation to the phase of the sexual cycle during which the call is emitted (Semple and McComb 2000). Playback experiments have shown that this variation is perceptually salient to males (Semple 1998; Semple and McComb 2000). Unlike rhesus macaques (Hauser 1993a; Hauser 1996), Barbary macaque males remain silent during copulation. For details, see Todt *et al.* 1995; Semple 1998; Semple, this volume.

Infant care

During interactions with infants, or while third parties interacted with an infant, Barbary macaques frequently uttered vocalizations. Figure 1a presents an example of a series of pants that were given while the subject was observing a triadic interaction involving an infant (Brumm *et al.* 2005). During so-called 'triadic interactions', also known as 'agonistic buffering' (Deag and Crook 1971; Deag 1980; Taub 1980a, b, 1984), subjects often produce "girneys", nasally sounding low amplitude calls. During triadic interactions, animals proceed through an elaborate greeting procedure that predominantly consists of retraction of the lips and rapid teeth chattering, while often also rapidly flapping the tongue with the mouth half open. Occasionally, animals also retract the scalp or shake their heads. This behaviour has also been characterised as 'smacking' (Paul 1984), and may grade into the 'silent bared-teeth display' (Preuschoft 1992). Girneys may also be heard when the group travels into another area of the park.

Quiet situation

Subjects may also vocalise in non-social contexts. Collectively, these calls may be described as 'soft calls' because they have a low amplitude, but considerable acoustic variation, ranging from soft grunts to short geckers and girneys. For the observer, the immediate cause of the calling is difficult to determine, and in the majority of instances, we simply had noted that the subject was sitting alone, with no particular activity visible or audible in the environment. Subjects also may vocalise when they are travelling from their day-time resting areas to their sleeping site. It does not seem likely that these calls are addressed to a particular recipient.

Agonistic situations

During agonistic interactions, Barbary macaques utter two main types of calls, namely screams and threat pants. Figure 1b presents call bouts given by a

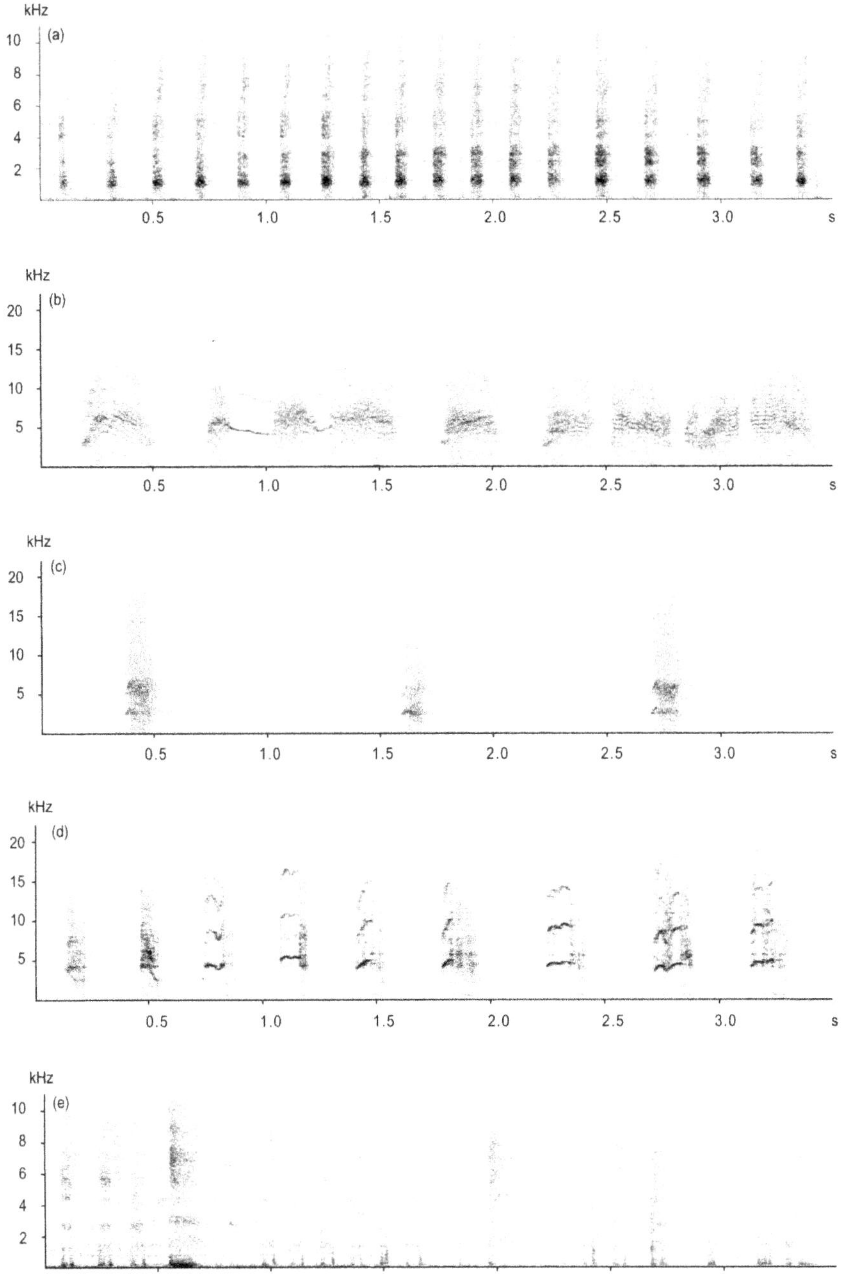

Figure 1 (a) Series of pants that were given while the subject was observing a triadic interaction involving an infant; (b) complex screams from a 3.5yr old male who challenged a female and appeared to recruit support from allies; (c) shrill barks recorded from an adult female after the animals had been disturbed on their sleeping trees at night; (d) squeaks and modulated tonal screams recorded from a yearling who attempted to establish contact with its mother during sleeping cluster formation at dusk; (e) vocalizations uttered during play consist of staccatos of soft aspirated low frequency grunts.

juvenile while she appeared to recruit allies for support (Gouzoules *et al.* 1984; Gouzoules and Gouzoules 1995). Some of these calls show insertions of a periodic window in the otherwise noisy (chaotic) structure of the call. While mild threat displays consist of a stare with the eyebrows slightly raised, more intense forms involve protruding the lips and forming a round opening often, but not always, accompanied by threat pants (Hesler and Fischer in press). These threat pants have a staccato-grunt structure, and the single units making up the call resemble the units of the mating call.

Disturbance

Figure 1c presents a series of shrill barks given in response to the appearance of a nearby dog (Fischer *et al.* 1995; Fischer 1998; Fischer and Hammerschmidt 2001). An acoustic analysis of this general call type revealed significant variation in relation to the stimulus that elicited the calling. Playback experiments have also shown that the subjects themselves were able to discriminate calls given in response to different stimuli (Fischer 1998; Fischer and Hammerschmidt 2001). Alarm calls given in response to birds are of lower frequency and have a rasping sound quality.

Sleeping cluster formation

Figure 1d presents calls recorded from a half-year old at dusk during sleeping cluster formation while it was trying to gain access to the same sleeping cluster as the mother, but was being rejected by her (Hammerschmidt *et al.* 1994). This calling has been viewed as an expression of the parent-offspring conflict (Trivers 1974).

Infant in care

When infants are in care of a male alloparent and attempt to get back to their mothers, they frequently emit long sequences of calls that can be classified as undulated screams (Hammerschmidt and Todt 1995). In many cases, males will return infants to their mother either when the mother attempts to retrieve the infant or when the infant attempts to return to its mother (Riechelmann *et al.* 1994). Some males, however, restrain the infant for long periods of time, causing considerable distress for the infant and the mother (Todt 1988; Riechelmann *et al.* 1994).

Play

Figure 1e presents vocalizations emitted during play, consisting of staccatos of soft aspirated low frequency grunts (Kipper and Todt 2002). During play, animals typically display an open-mouth round face (Preuschoft 1992; Hesler and Fischer 2005) while they vocalise.

Contact

Barbary macaques utter a variety of soft calls when they make or break contact; These typically can be described as short geckers.

Infant search

Females who are apparently looking for their infants, who are either in care of a male or that have been taken away by the park staff after the infant had died, emit tonal screams.

This qualitative description suggests that the vocal repertoire of Barbary macaques can be characterised as variations on a few themes, namely screams, shrill barks, geckers, and low frequency pants. Occasionally, Barbary macaques produce tonal 'coo'-like calls, and nasally sounding girneys. Similar sounding calls are given in different contexts, in turn, in some contexts, different call types can be observed. Intergradations within and between call types occur frequently.

Quantitative analysis

Corroborating the findings from the qualitative inspections, the quantitative analysis also failed to reveal clear cut call types. No single cluster solution appeared to be highly superior to all other solutions. Instead, the assessment of the different cluster solutions indicated that there were several possible solutions, in particular dividing the data set into 2-, 4-, 7- and 16-clusters (or call types). According to the measures reflecting the reduction of variance, we finally settled on the 7-cluster solution because it provided a balance of the cohesion within clusters and the isolation between clusters. In other words, the different clusters revealed noticeable differences among each other and sufficient similarity within the cluster.

Figure 2 shows the discrimination of the clusters by means of a discriminant function analysis. The discriminant function analysis is a tool to examine how well different groups (in this case: clusters) can be told apart. All clusters (except cluster G) could be clearly discriminated by two discriminant functions. These correlated with the distribution of the frequency in the spectrum and call duration, respectively. Spectrograms of calls which were assigned closest to the group centroid for each cluster, i.e. to the group mean, illustrate the call characteristics of each cluster (Figure 2b).

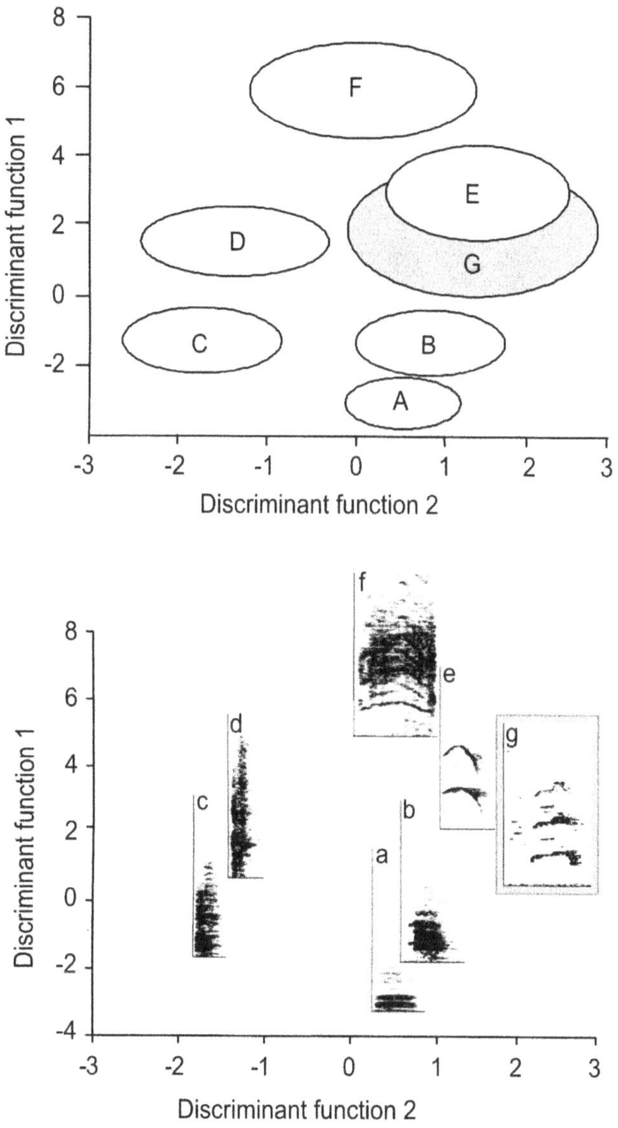

Figure 2 Arrangement of the clusters established by the 7-cluster solution in the acoustic space. (a) Circles are spanned by the x ± sd for the first two discriminant functions. Function 1 showed the highest correlation with variables describing the location of the maximum energy in the spectrum, while function 2 correlated best with variables describing the frequency range. (b) Spectrograms selected from the respective cluster centres, illustrating the acoustic characteristics of the call types.

Figure 3 shows the context-related usage of acoustic clusters. Since the different contexts were to some degree age- and sex-specific, the distribution of the use of the different call types corresponded to the one observed for the different age-classes (see below). For instance, calls recorded during sleeping

Call types									
	A	B	C	D	E	F	G	N calls	N subjects
mating (f)	●	●	●	○	○	○	○	1296	14
agonistic sit.	●	●	●	●	●	●	●	3286	66
disturbance	●	●	●	●	●	○	○	373	12
sleeping cluster	○	●	●	●	●	●	●	417	8
play	●	●	●	●	●	●	○	327	18

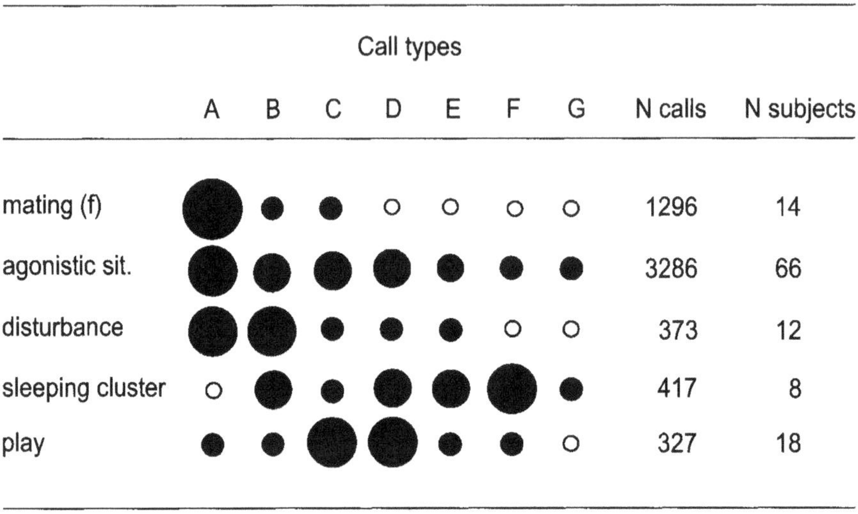

Figure 3 Call type usages in the different contexts. Circle sizes represent the percentage of call types given in the respective context: ○ = 0-<1%, ● = 1-<10%, 10-<30%, 30-<60%, and 60-100%. N calls = total number of calls per context, N subjects = number of subjects per context.

cluster formation were almost exclusively recorded from young juveniles. In this context, subjects mostly used calls of the clusters D, E, and F. In general, there was no simple relationship between context and the use of specific call types. All call types were observed in several contexts, and in all the contexts, several call types were used. In some contexts, for instance during copulation or when females were in search of their infants, there was a higher specificity in the call type usage. Splitting contexts into finer subcontexts did not change the result in a fundamental way.

The relative frequency of cluster usage revealed age-related differences. We observed a shift from a frequent use of scream-type calls and higher-pitched grunt-like calls in young animals to the use of low-frequency grunts in adults. Calls from infants and 1-year old animals were mainly assigned to call types D, E, and F. Juvenile calls typically belonged to call types C and D, with the exception that juvenile males also used calls from call type A. However, there were only few individuals represented in this age-class. Subadults and adults were mainly represented in call types A, B, and C. However, subadults were also represented in call types D and E. We found no difference in call usage between males and females. Also within clusters, we found significant age-related variation in a suite of acoustic variables, most notably in the mean central frequency, the mean local modulation of central frequency, mean 1st dominant frequency band, and mean local modulation of 1st dominant frequency band. Strikingly, we found only minor differences in relation to sex, namely in clusters A and C. Younger animals were found to produce calls with a higher local modulation of the central frequency and the 1st dominant frequency band, which could be related to a less well developed vocal motor control in younger animals. The mean central frequency and the 1st dominant frequency band decreased with increasing age. This decrease in pitch could be explained by an increase in body size and, correspondingly, the size of the larynx. Surprisingly, we did not find a significant difference between sexes in the present study despite a considerable sex dimorphism in body size (males are up to one third larger than females; Paul 1984). This finding is puzzling and still unexplained.

Discussion

The vocal repertoire of Barbary macaques may be described as an intergraded set of screams and shrill barks to geckers and low frequency pants. Occasionally, Barbary macaques produce tonal 'coo'-like calls, and nasally sounding girneys. Neither a qualitative assessment nor the quantitative analysis (Hammerschmidt and Fischer 1998) has led to the identification of distinct, highly stereotyped call types. Despite the large degree of intergradation, some acoustic patterns were observed more frequently than others. In our analysis, these emerged as 'call types' and represented the cluster centres.

As we illustrate in this present paper, there was no clear-cut relationship between call types and the situations in which they occurred. For instance, pant calls are given in agonistic encounters when one animal is threatening another, but also when a subject observes an interaction between group members and an infant, clearly an affiliative situation. Likewise, tonal calls were recorded from females that were threatened as well as from females who were in search of their infant. Although noisy and complex screams were most often observed in highly charged contexts such as contact aggression, it is important to keep in mind that this context category encompasses a wide variety of situations. Some call types, for instance noisy screams, are given by members of all age-sex-classes. The same is true for shrill barks, with the exception that infants do not produce them. Highly undulated screams, in contrast, are most frequently given by infant and juvenile macaques, and only rarely by members of older age classes (Hammerschmidt and Fischer 1998).

Although we cannot rule out the possibility that extended periods of observations of wild Barbary macaques would reveal additional calls or differences in call usage, we believe that our description covers the most important aspects of Barbary macaque vocalization. Also, Semple's description of the acoustic characteristics and usage of mating calls by Barbary macaques living in Gibraltar (Semple 1998; Semple and McComb 2000) and our personal observations in Gibraltar confirmed the view that our description reflects species-typical vocal behaviour.

It still remains difficult to compare the repertoires of different species as there is no standard method to classify call types. Also, the characterization of calls either according to presumed function or context may vary from study to study. To date, there are few comprehensive studies on the vocal repertoires of other macaques species, for instance Green's study on Japanese macaques, *M. fuscata* (Green 1975), Palombit's description of the vocal repertoire of long-tailed macaques, *M. fascicularis* (Palombit 1992), or Hohmann's overview of the calls of the lion-tailed macaques, *M. silenus* (Hohmann and Herzog 1985). Most other studies addressing macaque vocal behaviour dealt with more specific aspects, for instance the use of 'food calls' in toque macaques, *M. sinica*, (Dittus *et al.* 1991), acoustic variation within and between contexts (Hauser 1991; Gouzoules and Gouzoules 1995), or relation to age and sex (Green 1981; Inoue 1988; Hohmann 1991), and individual differences in vocalizations (Gouzoules and Gouzoules 1990; Rendall *et al.* 1996; Rendall *et al.* 1998). It is worth noting that in contrast to some other macaque species (Dittus *et al.* 1991; Hauser and Marler 1993), we never observed any vocalizations in the context of encountering food or highly preferred food. In sum, it seems that – despite the differences in the set-up and presentation of the above mentioned studies – Barbary macaques, the only African macaques, use considerably more harsh and noisy vocalizations and have a very limited use of 'coo calls' compared to the Asian macaques. This seems also to be true in comparison with another terrestrial macaque, the stump-tailed macaque, *M. arctoides* (Lillehei and Snowdon 1978).

The investigation of the structure of vocal repertoires of different species is particularly interesting in the light of the question of which factors underlie the diversity of different primate species vocal communication. Traditionally, repertoires have either been described as 'graded' or 'discrete', or a mixture of the two (Marler 1975; Marler 1976). A graded signal system is characterised by continuous acoustic variation between and/or within signal types, with no obvious distinct boundaries that allow a listener to discriminate easily between one signal type and another. Discrete repertoires, on the other hand, contain signals with no intermediates between call types. Marler hypothesised that graded vocal repertoires should evolve when individuals inhabit relatively open habitat and interact at high rates and at close range with conspecifics. In contrast, discrete vocal repertoires should be favoured when auditory signals must operate without accompanying visual or other contextual cues; for example, in forest habitats or when being broadcast over long distances

(Marler 1975; Marler 1976). At first glance, the Barbary macaques' repertoire fits with Marler's hypothesis. However, Marler also hypothesised that long-distance calls should be acoustically distinct because other cues may be lacking. With regard to the Barbary macaques' shrill barks, this certainly is not the case.

Currently, it seems that no single factor can account for the variation in repertoire diversity. Other factors such as body size, phylogenetic descent, and social structure presumably also play a role in shaping a species' repertoire (Hohmann and Herzog 1985; Hauser 1993b). As these factors are also related to one another it will be difficult to extract the primary driving forces. Nonetheless we believe that a systematic study of the diversity and variability of different nonhuman primate repertoires could provide important insights into the selective pressures and evolutionary constraints operating on the vocal behaviour of monkeys and apes.

Acknowledgements

We thank Ellen Merz and Gilbert de Turckheim for permission to conduct this study at Rocamadour, Keith Hodges and John Cortes for inviting us to contribute to this volume, and Viveka Ansorge and Dietmar Todt for crucial contributions in the early period of our Barbary macaque studies.

References

Bacher, J. 1994. Clusteranalyse. München: Oldenbourg Verlag.

Brumm, H., Kipper, S., Riechelmann, C. and Todt, D. 2005. Do Barbary macaques ‚comment' on what they see? A first report on vocalizations accompanying interactions of third parties. *Primates*, **46**: 141-144.

Darwin, C. 1872. The Expression of the Emotions in Man and Animals. London: Murray.

Deacon, T. W. 1992. The neuronal circuitry underlying primate calls and human language. In: *Language Origin: a Multidisciplinary Approach* (Ed. by J.Wind, B.Chiarelli, B.Bichakajian and A.Jonker), pp. 121-162.

Deag, J. M. 1980. Interactions between males and unweaned Barbary macaques: Testing the agonistic buffering hypothesis. *Behaviour*, **75**: 54-81.

Deag, J. M. and Crook, J. H. 1971. Social behaviour and agonistic buffering in the wild Barbary macaque, *Macaca sylvanus*. *Folia primatologica*, **15**: 183-200.

Dittus, W. P. J., Evans, C. S., Evans, L. and Marler, P. 1991. Toque macaque food calls: Semantic communication concerning food distribution in the environment. *Animal Behaviour*, **352**: 328-330.

Fischer, J. 2002. Developmental modifications in the vocal behaviour of nonhuman primates. In: *Primate Audition* (Ed. by A.A.Ghazanfar), pp. 109-125. Boca Raton: CRC Press.

Fischer, J. 1998. Barbary macaques categorize shrill barks into two call types. *Animal Behaviour*, **55**: 799-807.

Fischer, J. and Hammerschmidt, K. 2001. Functional referents and acoustic similarity revisited: The case of Barbary macaque alarm calls. *Animal Cognition*, **4**: 29-35.

Fischer, J., Hammerschmidt, K. and Todt, D. 1995. Factors affecting acoustic variation in Barbary macaque *(Macaca sylvanus)* disturbance calls. *Ethology*, **101**: 51-66.

Fischer, J., Kitchen, D. M., Seyfarth, R. M. and Cheney, D. L. 2004. Baboon loud calls advertise male quality: Acoustic features and their relation to rank, age, and exhaustion. *Behavioral Ecology and Sociobiology*, **56**: 140-148.

Fitch, W. T., Neubauer, J. and Herzel, H. 2002. Calls out of chaos: the adaptive significance of nonlinear phenomena in mammalian vocal production. *Animal Behaviour*, **63**: 407-418.

Geissmann, T. 2002. Duet-splitting and the evolution of gibbon songs. *Biological Reviews*, **77**: 57-76.

Ghazanfar, A. A. and Santos, L. R. 2004. Primate brains in the wild: The sensory bases for social interactions. *Nature Reviews Neuroscience*, **5**: 603-616.

Gouzoules, H. and Gouzoules, S. 1990. Matrilineal signatures in the recruitment screams of pigtail macaques, *Macaca nemestrina*. *Behaviour*, **115**: 327-347.

Gouzoules, H. and Gouzoules, S. 1995. Recruitment screams of pigtail monkeys (*Macaca nemestrina*) - ontogenetic perspectives. *Behaviour*, **132**: 431-450.

Gouzoules, H., Gouzoules, S. and Marler, P. 1984. Rhesus monkey (*Macaca mulatta*) screams: representational signalling in the recruitment of agonistic aid. *Animal Behaviour*, **32**: 182-193.

Green, K. P. 1975. Variation of vocal pattern with social situation in the Japanese monkey (*Macaca fuscata*): A field study. In: *Primate Behavior Vol.4* (Ed. by L.A.Rosenblum), pp. 1-102. New York: Academic Press.

Green, S. M. 1981. Sex differences and age gradations in vocalizations of Japanese and lion-tailed monkeys *(Macaca fuscata* and *Macaca silenus)*. *American Zoologist*, **21**: 165-183.

Hammerschmidt, K., Ansorge, V., Fischer, J. and Todt, D. 1994. Dusk calling in Barbary macaques (*Macaca sylvanus*): Demand for social shelter. *American Journal of Primatology*, **32**: 277-289.

Hammerschmidt, K. and Todt, D. 1995. Individual differences in the vocalizations of young Barbary macaques (*Macaca sylvanus*): a multiparametric analysis to identify critical cues in acoustic signalling. *Behaviour*, **132**: 381-399.

Hammerschmidt, K. and Fischer, J. 1998. The vocal repertoire of Barbary macaques: A quantitative analysis of a graded signal system. *Ethology*, **104**: 203-216.

Hauser, M. D. 1993b. The evolution of nonhuman primate vocalizations: Effects of phylogeny, body weight, and social context. *American Naturalist*, **142**: 528-542.

Hauser, M. D. 1996. *The Evolution of Communication. Cambridge*: MIT Press.

Hauser, M. D. 1991. Sources of acoustic variation in rhesus macaque (*Macaca mulatta*) vocalizations. *Ethology*, **89**: 29-46.

Hauser, M. D. 1993a. Rhesus monkey copulation calls: honest signals for female choice? *Proceedings of the Royal Society of London Series B-Biological Sciences*, **254**: 93-96.

Hauser, M. D. and Marler, P. 1993. Food-associated calls in rhesus macaques (*Macaca mulatta*). 1. Socioecological factors. *Behavioral Ecology*, **4**: 194-205.

Hesler, N. and Fischer, J. 2005. Gestures in Barbary macaques. In: *Gestural Communication in Monkeys and Apes* (Ed. by M.Tomasello and J.Call).

Hohmann, G. 1991. Comparative analyses of age-specific and sex-specific patterns of vocal behavior in 4 species of old-world monkeys. *Folia primatologica*, **56**: 133-156.

Hohmann, G. and Herzog, M. O. 1985. Vocal communication in lion-tailed macaques (*Macaca silenus*). *Folia primatologica*, **45**: 148-178.

Inoue, M. 1988. Age gradations in vocalizations and body weight in Japanese monkeys (*Macaca fuscata*). *Folia primatologica*, **51**: 76-86.

Jürgens, U. 1979. Neural control of vocalizations in nonhuman primates. In: *Neurobiology of Social Communication in Primates* (Ed. by H.D.Steklis and M.J.Raleigh), pp. 11-44. New York: Academic Press.

Kipper, S. and Todt, D. 2002. The use of vocal signals in the social play of Barbary macaques. *Primates*, **43**: 3-17.

Kitchen, D. M., Fischer, J., Cheney, D. L. and Seyfarth, R. M. 2003. Loud calls as indicators of dominance in male baboons (*Papio cynocephalus ursinus*). *Behavioral Ecology and Sociobiology*, **53**: 374-384.

Lillehei, R. and Snowdon, C. T. 1978. Individual and populational differences in the vocalizations of young stumptail macaques (*Macaca arctoides*). *Behaviour*, **65**: 270-281.

Marler, P. 1976. Social organization, communication and graded signals: The chimpanzee and the gorilla. In: *Growing Points in Ethology* (Ed. by P.P.G.Bateson and R.A.Hinde), pp. 239-280. Cambridge UK: Cambridge University Press.

Marler, P. 1975. On the origin of speech from animal sounds. In: *The Role of Speech in Language* (Ed. by J.F.Kavanaugh and J.E.Cutting), pp. 11-37. Cambridge, MA: MIT Press.

Palombit, R. A. 1992. A preliminary study of vocal communication in wild long-tailed macaques (*Macaca fascicularis*). 1. Vocal repertoire and call emission. *International Journal of Primatology*, **13**: 143-182.

Paul, A. 1984. Zur Sozialstruktur Und Sozialisation Semi-Freilebender Berberaffen (*Macaca Sylvanus* L.1758). Dissertation. Göttingen: Univ. Göttingen.

Paul, A., Küster, J. and Arnemann, J. 1996. The sociobiology of male-infant interactions in Barbary macaques (*Macaca sylvanus*). *Animal Behaviour*, **51**: 155-170.

Preuschoft, S. 1992. "Laughter" and "Smile" in Barbary macaques (*Macaca sylvanus*). *Zeitschrift für Tierpsychologie*, **91**: 220-236.

Rauschecker, J. P., Tian, B. and Hauser, M. D. 1995. Processing of complex sounds in the macaque nonprimary auditory cortex. *Science*, **268**: 111-114.

Rendall, D., Owren, M. J. and Rodman, P. S. 1998. The role of vocal tract filtering in identity cueing in rhesus monkey (*Macaca mulatta*) vocalizations. *Journal of the Acoustical Society of America*, **103**: 602-614.

Rendall, D., Rodman, P. S. and Emond, R. E. 1996. Vocal recognition of individuals and kin in free ranging rhesus monkeys. *Animal Behaviour*, **51**: 1007-1015.

Riechelmann, C., Hultsch, H. and Todt, D. 1994. Early development of social relationships in Barbary macaques (*Macaca sylvanus*): Trajectories of alloparental behaviour during an infant's first three months of life. In: *Current Primatology* (Ed. by J.J.Roeder, B.Thierry, J.R.Anderson and N.Herrenschmidt), pp. 279-268. Strasbourg: Universite Louis Pasteur.

Semple, S. 1998. The function of Barbary macaque copulation calls. *Proceedings of the Royal Society of London Series B-Biological Sciences*, **265**: 287-291.

Semple, S. and McComb, K. 2000. Perception of female reproductive state from vocal cues in a mammal species. *Proceedings of the Royal Society of London Series B-Biological Sciences*, **267**: 707-712.

Semple, S., McComb, K., Alberts, S. and Altmann, J. 2002. Information content of female copulation calls in yellow baboons. *American Journal of Primatology*, **56**: 43-56.

Seyfarth, R. M. and Cheney, D. L. 2003. Signalers and receivers in animal communication. *Annual Review of Psychology*, **54**: 145-173.

Small, M. F. 1990. Promiscuity in Barbary macaques (*Macaca sylvanus*). *American Journal of Primatology*, **20**: 267-282.

Taub, D. M. 1980a. Female choice and mating strategies among wild Barbary macaques (*Macaca sylvanus*). In: *The Macaques: Studies in Ecology,*

Behaviour and Evolution (Ed. by D.M.Taub) New York: Van Nostrand Reinhold.

Taub, D. M. 1984. Male caretaking behaviour among wild Barbary macaques. In: *Primate Paternalism* New York: Van Nostrand Reinhold.

Taub, D. M. 1980b. Testing the 'agonistic buffering' hypothesis. 1. The dynamics of participation in the triadic interaction. *Behavioral Ecology and Sociobiology*, **6**: 187-197.

Tinbergen, N. 1963. On aims and methods in ethology. *Zeitschrift für Tierpsychologie*, **20**: 410-433.

Todt, D. 1986. Hinweis-Charakter und Mittler-Funktion von Verhalten. *Z Semiotik*, **8**: 183-232.

Todt, D. 1988. Serial calling as a mediator of interaction processes: Crying in primates. In: *Primate Vocal Communication* (Ed. by D.Todt, P.Goedeking and D.Symmes) Berlin: Springer.

Todt, D., Hammerschmidt, K., Ansorge, V. and Fischer, J. 1995. The vocal behaviour of Barbary macaques: Call features and their performance in infants and adults. In: *Current Topics in Primate Vocal Communication* (Ed. by E.Zimmermann, J.D.Newman and U.Jürgens), pp. 141-160. New York: Plenum Press.

Trivers, R. L. 1974. Parent-Offspring Conflict. *American Zoologist,* **14**: 249-264.

Turckheim, G. D. and Merz, E. 1984. Breeding Barbary macaques in outdoor open enclosures. In: *The Barbary Macaque: A Case Study in Conservation* (Ed. by J.F.Fa), pp. 241-261. New York: Plenum Press.

The function of female copulation calls in the genus *Macaca*: insights from the Barbary macaque

S Semple[1] and K McComb[2]
[1]*Centre for Research in Evolutionary Anthropology, School of Human and Life Sciences, Roehampton University, London SW15 4JD, UK;*
[2]*Department of Psychology, School of Life Sciences, University of Sussex, Brighton BN1 9QH, UK*

Introduction

In the genus *Macaca*, sex is often a noisy affair. Mating in many macaque species, including the Barbary macaque, is accompanied by the loud, rhythmic copulation calls of females (Figure 1, Table 1). While many animal species produce distinctive vocalisations around the time of mating, the female copulation calls of macaques have generated particular interest not only among primatologists, but also among behavioural ecologists more widely, for two main reasons. Firstly, the fact that it is the females that call is unusual, as vocal signals given in mating contexts are typically given by males. Secondly, the timing of the call - typically at the end of the copulation - also merits attention as vocalisations given in mating contexts usually precede rather than follow mating. Female copulation calls in macaques therefore seem something of an oddity. Moreover, the loud volume and complex acoustic structure of this signal call strongly suggests it has a signalling function rather than being a non-adaptive by-product of mating behaviour (Hamilton and Arrowood, 1978).

To date, a large number of hypotheses have been put forward to explain the benefits that females might gain from calling (for a comprehensive discussion of these, see review by Maestripieri and Roney, 2005). Particular attention has been paid to those hypotheses in which the call is proposed to affect the behaviour of either the mating male, or other males within the group. These suggest that female copulation calls may play an important role in shaping patterns of male sexual behaviour and consequently may underpin unusual female sexual strategies. These hypotheses are also the only ones to be tested so far among the macaques and consequently they form the focus for this chapter. While evidence is lacking for other hypotheses - for example those in which the receivers are proposed to be other females

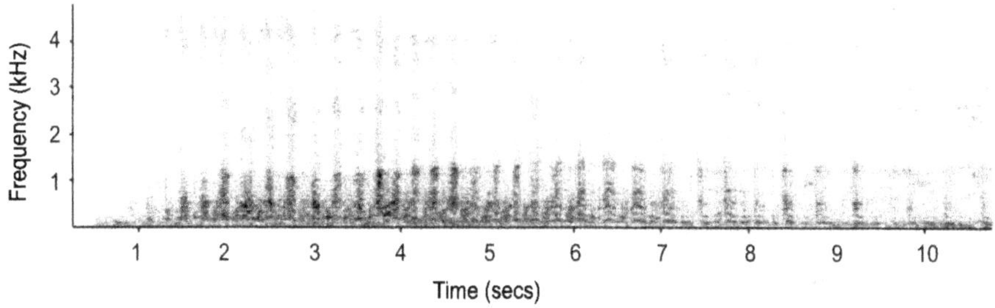

Figure 1. Spectrogram of a female Barbary macaque copulation call.

Table 1. Occurrence of female copulation calls and prominent sexual swellings in the genus *Macaca.*

Species	*Call*	*Prominent swelling*[1]	*Reference for copulation call*
Macaca arctoides	absent	absent	Bertrand (1969)
Macaca cyclopis	present	present	Hsu *et al.* (2005)
Macaca fascicularis	present	present	Deputte and Goustard (1980)
Macaca fuscata	present	present	Oda and Masataka (1992)
Macaca mulatta	absent	present	Hauser (1996)
Macaca nemestrina	present	present	van Schaik *et al.* (1999)
Macaca nigra	present	present	O Petit (personal communication)
Macaca radiata	present	absent	Hohmann (1989)
Macaca silenus	present	present	Hohmann (1991)
Macaca sylvanus	present	present	Pohl and Todt (1984)
Macaca thibetana	present	absent	van Schaik *et al.* (1999)
Macaca tonkeanna	present	present	van Schaik *et al.* (1999)

[1]Data taken from Nunn (1999)

in the group (Hohmann and Herzog,1985) - it should be noted that the right sort of data to test these hypotheses has simply not yet been collected. Below we examine the hypothesised benefits to females of copulation calling and evaluate the evidence from macaques relating to these hypotheses.

Copulation calls function to incite males to compete for access to the calling female

Females would be predicted to benefit from their copulation calls inciting males to compete for access to them, as fitter males are more likely to emerge victorious from such competitive interactions. In this way, copulation calls would allow females to select indirectly for the 'best' males in the group without paying the potential costs of mate searching or assessment (Wiley and Poston, 1996). There are at least three different time scales over which the results of male-male competition might be manifested; the benefits that females stand to gain under these three scenarios differ in important ways (Semple, 1998a).

Females could, in theory, incite males to interfere in the very copulation in which the call is given. If calling led to displacement of the mating male prior to ejaculation, only the strongest males would successfully inseminate the female, as these animals alone would be able to prevent displacement by other group males. In this way, calls would allow females indirectly to select for the strongest male(s) as potential father(s) to their future infants. At first sight, this may seem an unlikely scenario among macaques, as females of many species call right at the end of the mount, just before or even after ejaculation occurs - too late for a male responding to the call to displace the mounting male and prevent insemination. However, this hypothesis could be relevant to multiple mount to ejaculation species if calls are given during or following an early mount in the series. Working with one such species, the Japanese macaque, Oda and Masataka (1995) found that interference in mount sequences was significantly more likely when a copulation call occurred than when copulations were 'silent'. On the surface, this evidence seems to fit rather nicely with the hypothesised benefits of calling. However, a range of other factors such as female reproductive state may also affect both likelihood of calling and the attractiveness of females to males (and hence males' willingness to compete for them). The reported relationship between calling and probability of interference may therefore potentially be an artefact, resulting from a lack of control for confounding variables.

For single mount to ejaculation macaque species, such as the bonnet macaque (Shively *et al.*, 1982) and stumptail macaque (Chevalier-Skolnikoff, 1975), interference in copulations is very rare indeed. In these species, copulation calling may incite competition among males for the copulation subsequent to the one in which the call is given (Semple, 1998a). A rank-related difference in male attraction to female copulation calls (i.e. higher ranked males respond more strongly to this vocalisation) would also result in indirect mate choice and consequently calling females would, on average, be mated by higher ranked males. By reducing the intervals between copulations and also encouraging a greater number of males to mate with her, a calling female could also benefit from the resulting sperm competition between males. This would lead to many or all males having a non-zero probability of fathering the female's infant; this shared probability of paternity could prove an effective strategy against infanticide (Hrdy, 1979). Another potential benefit to females of inciting sperm competition would accrue if sperm quality is heritable, since females may benefit through the increased reproductive success of their sons (Möller and Birkhead, 1989).

Copulation calls may function to alert the dominant male in the group to the female's probability of conceiving (Henzi, 1996). As a result, this male would consort with the female during her most fertile period, and exclude the mating attempts of other males. Effectively, the female by calling is indirectly selecting for the dominant male as most likely father to her infant, and would consequently benefit from the care and protection given to the infant by this powerful male. This hypothesis was originally formulated to explain the function of female copulation calls in chacma baboons, a species in which consortships are long and the alpha male generally has sole access to females during the period that they are likely to conceive (Henzi, 1996). Among macaque species, it is relevant only to those where such monopolisation of females is found, such as the Tonkean macaque (Aujard *et al.*, 1998) and lion-tailed macaque (Kumar and Kurup, 1985). The highly seasonal nature of breeding in this genus means that monopolisation of this sort is by no means ubiquitous. A study of the function of female copulation calls in a captive group of long-tailed macaques set out to test the idea that these calls signal information about likelihood of conception to the dominant male and encourage his establishment of consortship (Nikitopoloulos *et al.*, 2004). In support of this hypothesis, it was found that females mated mainly with the alpha male during the period when conception occurred. Crucially, however, the exact role (if any) that copulation calls play in producing this pattern of mating behaviour was not examined directly and remains unclear.

Copulation calls function to incite mate guarding from mating males

An intriguing new hypothesis about the benefits to females of calling has recently been put forward by Maestripieri and Roney (2005), who propose that females' calls are primarily directed at the mating male. Females often turn their heads to look directly at the mating male while they produce the call, providing some indication that he may indeed be an intended receiver. Maestripieri and Roney (2005) suggest that copulation calls encourage mate guarding from preferred males, with calls predicted to be absent from copulations with unpreferred males. We would modify this prediction somewhat to suggest that calls would be predicted to vary in form (not just occurrence) according to the quality of the mating male. Interestingly, this hypothesis proposes that calls also incite other males to compete for access to the calling female. The authors state that "By using vocalisations that can also be heard by other males, however, females can also effectively blackmail their consort partners and further encourage them to engage in mate guarding behaviour".

Support for this hypothesis comes from one macaque species, the Japanese macaque, in which it has been reported that the likelihood of a female calling during copulation is higher during mating with adult than subadult males (Oda and Masataka, 1992). However, as a number of factors other than male age may affect whether or not a call is given during copulation, a lack of control for confounding variables in the analysis means this result should be treated with caution. It is possible, for example, that adult males are more likely to mate with females during their most likely period of conception and hence the apparent relationship between male age and occurrence of calls could be the result of a relationship between likelihood of calling and female reproductive state. Overall, support for this hypothesis among the macaques is lacking but a study on yellow baboons demonstrated that the rank of the mating male has a direct effect on the form of female copulation calls in this species (Semple *et al.*, 2002). After controlling for the effect of confounding variables, calls were found to be longer and contained more units during matings with higher ranked males. Only future research will indicate whether female macaques' copulation calls vary in a similar way and, if so, how this variation affects the behaviour of mating males.

The function of female copulation calls in Barbary macaques

Working with a study population in Gibraltar, we carried out playback experiments and acoustic analysis to examine the function and information content of female copulation calls in Barbary macaques. In a first series of experiments we tested the hypotheses that calling may benefit females by increasing both the quality of males with which they mate and the number of copulations they receive (Semple, 1998b). In a final experiment we tested whether males can assess female reproductive state from copulation calls (Semple and McComb, 2000).

Do female copulation calls lead to matings with higher quality males?

We first assessed whether female copulation calls provide females with a mechanism of indirect mate choice, resulting in their being mated by higher ranked males in the group. Of the three possible time scales over which male-male competition for access to the calling female might be incited (see discussion above), two have little relevance to the Barbary macaque. It seems very unlikely, firstly, that calls provide females with indirect mate choice for the immediate copulation as interruption preventing ejaculation by the mating male is extremely rare (occurring in less than 1% of copulations – Semple, 1998a). This is at least in part due to the timing of the call in this species: females call from shortly before dismount of the male until after ejaculation and dismount. Any male attending to the signal would in most cases be too late to prevent a successful insemination of the female. It seems equally unlikely that copulation calls lead to monpolisation of the female by the alpha male during her likely period of conception. The Barbary macaque has a 'promiscuous' mating system (Small, 1990) characterised by brief consortships, with females rarely mating twice in succession with the same male (Taub, 1982; Küster and Paul, 1992).

We therefore considered the remaining type of male-male competition that copulation calls might incite – access for the subsequent copulation. The hypothesis in this case would predict that males would be attracted to calling females and, crucially, that high ranked males would respond more strongly than low ranked individuals. Todt and Pohl (1984) found that male Barbary macaques did indeed approach loudspeakers from which female copulation calls were played, suggesting they are attracted to this signal, but the authors did not investigate any effect of male rank on response to the copulation call. Investigating possible rank effects on males' response to this signal was the aim of the first two playback experiments that we carried out.

In the first experiment, female copulation calls were played to males who were at least 20m away from the nearest other male. The effect of males' rank on direct response to this signal was assessed by comparing responses of high ranked (in the top half of the hierarchy) and low ranked (bottom half of the hierarchy) males. All males looked toward the loudspeaker on playback of the copulation call, and some subsequently approached it. High-ranking males were, however, no more likely to approach the loudspeaker than low-ranking males: 7 out of 12 high ranked males approached the loudspeaker, and 8 out of 12 low ranked males did so (Likelihood ratio test : $C^2=0.793$, df=1, NS). At face value, this result seemed not to support the hypothesis that copulation calls function to allow females indirect mate choice for the next copulation. Males were interested in calls and approached them, as predicted by this hypothesis, but there were no apparent rank effects on response.

However, the scenario in which males were tested (distant from other males) is an unusual one. More commonly males hearing female copulation calls have a number of other males in close proximity. Our second playback experiment therefore assessed whether male rank relative to neighbours (rather than male rank *per se*) might have an impact on how males responded to the copulation calls of females. Here, playbacks were made not to singletons but to pairs of males sitting in full view of each other and less than 10m apart. Unlike in the previous experiment where the rank of males was absolute, here rank was relative i.e. males in a pair were designated as dominant and subordinate. A number of males appeared as dominant in one trial and subordinate in another. In this experiment, a rank based difference in response was clear. In none of the 14 trials did a subordinate in the pair make an approach to the loudspeaker, while in 7 trials the dominant animal of the pair approached (Likelihood ratio test : $C^2=12.08$, $df=1$, $P<0.001$). Having established an effect of rank on the approach response of males, we then went on to test whether females, by calling, affect not just the quality but the number of their mating partners.

Do female copulation calls lead to more matings?

We tested the specific hypothesis that female copulation calls increase the number of copulations a female receives and hence promote sperm competition. Calls could produce this effect by reducing inter-copulation intervals as consort turnovers are frequent in Barbary macaques and there is a highly significant correlation between the number of total copulations and the number of different male partners that a female has i.e. females which mate more frequently do so with more partners (Small, 1990).

When a chosen receptive female was out of sight of all subadult and adult males in the troop, and at least 10m from the nearest one, her copulation call or a control stimulus (white noise) was played from a concealed loudspeaker within 5m of her. The following day in the same scenario the other of the two stimuli was played, giving a matched pair of presentations for each female. Following each playback, females were followed and the time to their subsequent copulation recorded. Ten different females were used as subjects in this experiment. Comparison within pairs revealed that females were mated significantly more quickly if their call had been played compared to when only the control stimulus was presented (Wilcoxon matched pairs signed-ranks test $T=3$, $n=10$ pairs, $P<0.01$; see Figure 2).

Taken together, the results of these three experiments indicate that by copulation calling, female Barbary macaques may increase both the number and quality of their mating partners. As such, they provide evidence that female copulatory vocalisations in this species are sexually selected traits. In itself, this is an interesting result: behavioural ecologists are used to examining sexually selected vocal signals but typically these are calls produced by males. There is as yet very little evidence that females may vocalise to attract mates (for exceptions see: Langmore *et al.*, 1996; Langmore and Davies, 1997). As far as we are aware, our findings represented the first such experimental evidence for primates.

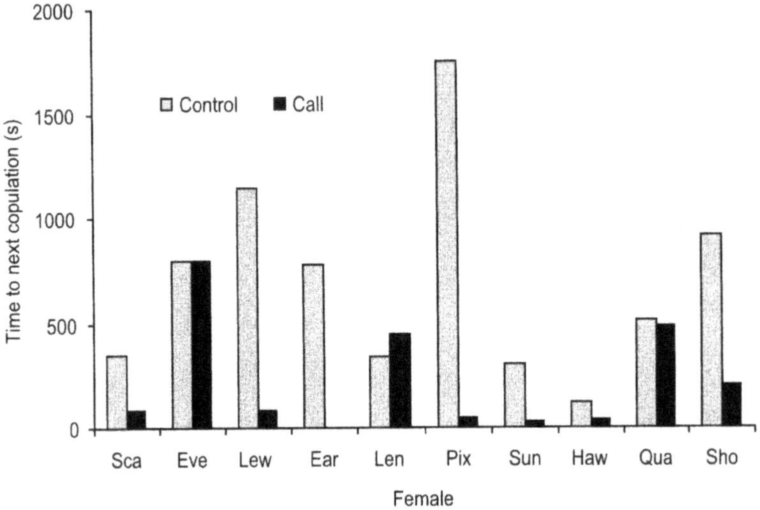

Figure 2. Time in seconds to copulation of females following playback of their copulation call or a control stimulus.

The notion that the evolution of Barbary macaque female copulation calls has been driven by sexual selection led us to ask one further question: do these signals indicate some aspect of the quality of the signaller? This is the sort of question that studies of male sexually selected vocal signals routinely address; in doing so, they generally seek evidence for inter-individual differences in some trait that the choosing sex might be selected to favour (Snowdon, 2004). But information may also be encoded in calls on within-individual temporal changes in 'quality'. One such aspect of female quality that males might be expected to attend to would be their reproductive state. Given the promiscuous nature of the mating system of Barbary macaques and the energetic demands on males during the mating season (Small, 1990), they would be predicted to allocate their mating efforts judiciously, with females becoming more attractive as mating partners when they are most likely to conceive. This prediction was tested in our final study.

Can males assess female reproductive state from their copulation calls?

In our final experiment, males were presented with playbacks of two copulation calls from the same female – one recorded early in the ovarian cycle and the other recorded at or near peak receptivity (as assessed by the size of the sexual swelling). These two playbacks occurred one to two days apart. Possible differences between calls from a particular female that were not related to female reproductive state were controlled for by only using calls from ejaculatory copulations with males of the same age group (adult/subadult) (see Semple and McComb, 2000 for further details on experimental protocols). When the response of each male to the playback pairs was compared a very clear pattern emerged: 11 of the 12 male subjects responded more strongly to the call given during peak receptivity (Sign test x=1, P<0.01). Subsequent analysis of the calls used in this experiment revealed that calls during peak receptivity (i.e. closer to the time of ovulation) were significantly longer, and contained more calls units of greater duration and higher mean dominant frequency but these were delivered at a lower rate (Wilcoxon matched-pairs signed ranks test n=12, P<0.05 in all cases). These results indicate that the copulation calls of female Barbary macaques signal temporal changes in 'quality' (female reproductive state) and that males are able to use such calls in order to assess the reproductive state of calling females.

Comparative perspectives

The results of our studies on female copulation calls in the Barbary macaque may have broader implications for other species in the genus. They suggest, first of all, that it may be profitable to investigate potential sexually selected benefits of female copulation calls among other macaque species. The idea that female macaques may be able to dramatically modify male sexual behaviour through their vocal communication suggests that our understanding of the mechanisms of sexual selection in this genus may be lacking some key insights. Studies of the way in which sexual selection may shape primate vocal signals have, until quite recently, been focussed almost exclusively on those vocalisations produced by males. Future theoretical and empirical studies of mate choice and sexual selection among macaques should therefore consider selective mechanisms from the perspective of males and females alike (Snowdon, 2004).

Our finding that males perceive female reproductive state from copulation calls in the Barbary macaque should also prompt a rethink of theories related to the evolution of concealed ovulation. The secondary loss of a prominent sexual swelling seen in a number of primate taxa including the genus *Macaca* (Dixson, 1983; Nunn, 1999) has led some authors to hypothesise on the evolutionary benefits of concealing (e.g. Andelamn, 1987; Sillen-Tullberg and Moller, 1993) or simply not widely advertising (Burt, 1992) information about reproductive state. Our results suggest that this information may potentially be broadcast over long distances in the acoustic rather than the visual modality. In a number of macaque species that lack prominent sexual swellings, females give copulation calls (Table 1). It seems premature to conclude that, in these species, female reproductive state is concealed from distant conspecifics. Important information may instead be coded in the copulation call, whose loud volume ensures it is broadcast over significant distances to a large number of potential receivers.

Finally, we suggest that the macaque genus provides an ideal group in which to adopt a comparative approach to the study of the evolution of female copulation calls. Such an approach, however, will require a greater resolution of data than has previously been available. Information on whether calls are simply present or absent within a species is not sufficient. Female copulation calls among macaques vary not just in whether they are given or not but also, critically, in their temporal and acoustic structure. This variation in call form may be related to identity of the caller (Barbary macaque: Todt *et al.*, 1995; long-tailed macaque: Deputte and Goustard, 1980), her reproductive state (Barbary macaque: Semple and McComb, 2000), whether ejaculation occurs (long-tailed macaue: Deputte and Goustard, 1980), and the age of the mating male (Japanese macaque: Oda and Masataka, 1995). Only when such variation has been quantified across a number of macaque species will a rigorous comparative analysis have the potential to test hypotheses relating to the evolutionary benefits of this signal. The growing number of ongoing studies examining the nature and functional significance of variation in the form of female macaque copulation calls is therefore encouraging.

References

Andleman SJ (1987) Evolution of concealed ovulation in vervet monkeys (*Cercopithecus aethiops*). *American Naturalist* **129** 785-799

Aujard F, Heistermann M, Thierry B and Hodges JK (1998) Functional significance of behavioral, morphological, and endocrine correlates across the ovarian cycle in semifree ranging female Tonkean macaques. *American Journal of Primatology* **46** 285-309

Bertrand M (1969) *The behavioural repertoire of the stumptail macaque*. Karger, New York

Burt A (1992) "Concealed ovulation" and sexual signals in primates. *Folia Primatologica* **58** 1-6

Chevalier-Skolnikoff S (1975) Heterosexual copulatory patterns in stumptail macaques (Macaca arctoides) and in other macaque species. *Archives of Sexual Behavior* **4** 199-200

Deputte BL and Goustard M (1980) Copulatory vocalisations of female macaques (*Macaca fascicularis*) : variability factors analysis. *Primates* **21** 83-99

Dixson AF (1983) Observations on the evolution and behavioural significance of "sexual skin" in female primates. *Advances in the Study of Behaviour* **13** 63-106

Hamilton WJ and Arrowood PC (1978) Copulatory vocalisations of chacma baboons (*Papio ursinus*), gibbons (*Hylobates hoolock*), and humans. *Science* **200** 1405-1409

Hauser MD (1996) *The Evolution of Communication.* MIT Press, Cambridge

Henzi SP (1996) Copulation calls and paternity in chacma baboons. *Animal Behaviour* **51** 233-234

Hohmann G (1989) Vocal communication of wild bonnet macaques (*Macaca radiata*). *Primates* **30** 325-345

Hohmann G (1991) Comparative analyses of age- and sex-specific patterns of vocal behaviour in four species of Old World monkeys. *Folia Primatologica* **56** 133-156

Hohmann G and Herzog MO (1985) Vocal communication in lion-tailed macaques (*Macaca silenus*). *Folia primatologica* **45** 148-178

Hrdy SB (1979) Infanticide among animals: a review, classification and examination of implications for the reproductive strategies of females. *Ethology and Sociobiology* **1** 13-40

Hsu MJ, Chen L-M, and Agoramoorthy G (2005) The vocal repertoire of Formosan macaques, *Macaca cyclopis*: acoustic structure and behavioural context. *Zoological studies* **44** 275-294

Küster J and Paul A (1992) Influence of male competition and female mate choice on male mating success in Barbary macaques (*Macaca sylvanus*). *Behaviour* **120** 192-217

Kumar A and Kurup GU (1985) Sexual behavior of the lion-tailed macaque, *Macaca silenus*. In *The lion-tailed macaque: status and conservation* pp 109-130 Ed PG Heltne. Alan Liss, New York

Langmore NE and Davies NB (1997) Female dunnocks use vocalizations to compete for males. *Animal Behaviour* **53** 881-890

Langmore NE, Davies NB, Hatchwell BJ and Hartley IR (1996) Female song attracts males in the alpine accentor *Prunella collaris. Proceedings of the Royal Society of London, Series B* **263** 141-146

Maestripieri D and Roney JR (2005) Primate copulation calls and postcopulatory female choice. *Behavioural Ecology* **16** 106-113

Möller AP and Birkhead TR (1989) Copulation behaviour in mammals : evidence that sperm competition is widespread. *Biological Journal of the Linnaean Society* 38, 119-131

Nikitopoulos E, Arnhem E, van Hooff JARAM, Sterck EHM (2004) Influence of female copulation calls on male sexual behavior in captive *Macaca fascicularis. International Journal of Primatology* **25** 659-677

Nunn CL (1999) The evolution of exaggerated sexual swellings in primates and the graded signal hypothesis. *Animal Behaviour* **58** 229-246

O'Connell SM and Cowlishaw G (1994) Infanticide avoidance, sperm competition and mate choice : the function of copulation calls in female baboons.

Animal Behaviour **48** 687-694

Oda R and Masataka N (1992) Functional significance of female Japanese macaque copulatory calls. *Folia primatologica* **58** 146-149

Oda R and Masataka N (1995) Function of copulatory vocalizations in mate choice by females of Japanese macaques (*Macaca fuscata*). *Folia Primatologica* **64** 132-139

Pohl R and Todt D (1984) Significance of mating calls and gestures in the Barbary macaque and their relation to the estrous cycle of females (*Macaca sylvanus* L.). *International Journal of Primatology* **5** 373

Semple S, McComb K, Alberts SC and Altmann J (2002) Information content of female copulation calls in yellow baboons. *American Journal of Primatology* **56** 43-56

Semple S (2001) Individuality and recognition of copulation calls of yellow baboons (*Papio c. cynocephalus*). *Animal Behaviour* **61** 1023-1028

Semple S and McComb K (2000) Perception of female reproductive state from vocal cues in a mammal species. *Proceedings of the Royal Society of London, Series B* **267** 707-712

Semple S (1998a) *Female copulation calls in primates*. D.Phil. University of Sussex.

Semple S (1998b) The function of copulation calls in the Barbary macaque (*Macaca sylvanus*). *Proceedings of the Royal Society of London, Series B* **265** 287-291

Shiveley C, Clarke S, King N, Schapiro S and Mitchell G (1982) Patterns of sexual behaviour in male macaques. *American Journal of Primatology* **2** 373-384

Sillen-Tullberg B and Moller AP (1993) The relationship between concealed ovulation and mating systems in anthropoid primates - a phylogenetic analysis. *American Naturalist* **141** 1-25

Small MF (1990) Promiscuity in Barbary macaques (*Macaca sylvanus*). *American Journal of Primatology* **20** 267-282

Snowdon CT (2004) Sexual selection and communication. In *Sexual selection in primates – new and comparative perspectives* pp57-70 Eds P Kappeler and C van Schaik. Cambridge University Press, Cambridge

Taub DM (1982) Sexual behaviour of wild Barbary macaque males (*Macaca sylvanus*). *American Journal of Primatology* **2** 109-113

Todt D and Pohl R (1984) Communicative strategies in estrous Barbary ape females (*Macaca sylvanus* L.) during copulation behaviour : advertising, triggering, affiliating. *Verh. Dtsch. Zool. Ges.* **77** 225

Todt D, Hammerschmidt K, Ansorge V and Fischer J (1995) The vocal behaviour of Barbary macaques (*Macaca sylvanus*) : call features and their performance in infants and adults. In *Current topics in primate vocal communication* pp 141-160 Eds E Zimmerman, JD Newman and U Jurgens.

Plenum Press, New York
Van Schaik CP, van Noordwijk MA and Nunn CL (1999) Sex and social evolution in primates. In *Comparative Primate Socioecology* pp204-231 Ed PC Lee. Cambridge University Press, Cambridge
Wiley RH and Poston J (1996) Indirect mate choice, competition for mates, and coevolution of the sexes. *Evolution* **50** 1371-1381

The male "Grispol" copulates with "Grande", an occasional sexual partner. Copyright D. Vallet

Reproductive cycles and mating patterns in female Barbary macaques

M Heistermann[1], U Möhle[1], J Dittami[2], B Wallner[3], E Shaw[4], JK Hodges[1]
[1]*Department of Reproductive Biology, German Primate Centre, Kellnerweg 4, 37077 Göttingen, Germany*
[2]*Department of Behavioural Biology and Neuroscience, University of Vienna, Althanstrasse 14, 1090 Vienna, Austria*
[3]*Department of Anthropology, University of Vienna, Althanstrasse 14, 1090 Vienna, Austria*
[4]*Gibraltar Ornithological and Natural History Society, Gibraltar Natural History Field Centre, Upper Rock Nature Reserve, Gibraltar*

Reproductive seasonality

As an adaptation to the climatic conditions in its natural range, the Barbary macaque shows a marked degree of seasonality in breeding, characterised by the presence of a distinct mating season in autumn and winter and a short and discrete birth season timed to spring and early summer (Küster and Paul, 1984; Mehlman, 1989; Menard and Vallet, 1993). Thus, in common with three other species of its genus living in a marked temperate environment (*M. radiata, M. fuscata, M. cyclopis*; see Bercovitch and Harvey, 2004), the Barbary macaque exhibits highly pronounced reproductive seasonality. In contrast, the majority of macaques living in more tropical latitudes are either only moderately seasonal (as indicated by seasonal birth peaks, e.g. *M. fascicularis; M. maurus*) or breed throughout the year (e.g. *M. arctoides; M. silenus*; see Bercovitch and Harvey, 2004).

Behavioural studies have shown that the mating season in the Barbary macaque is characterised by the onset of regular copulations in September and early October and by a 2-3 month period of intense mating activity between October and January (Küster and Paul, 1984). Furthermore, timing of conceptions within this period appears to be highly synchronised between females. Küster and Paul (1984), for example, reported that almost 60% of the females located at Affenberg Salem, Germany, conceived within one month, between mid November and mid December. We found an even more pronounced synchrony in timing of conceptions in females of the Gibraltar population, where all of the conceptions recorded on the basis of endocrine information (n=21) during the three mating seasons 2001/2002, 2003/2004 and 2004/2005 consistently fell within a 5 week period (December 7 - January 13). Moreover, as reported for seasonally breeding rhesus monkeys (Lindburg, 1971) and Japanese macaques (Takahata, 1980), most conceptions occur in the first cycle of the mating season (Möhle *et al.*, 2005). Although data on timing of conceptions from wild Barbary macaque populations are lacking, a high degree of synchrony in timing of births in females living in Morocco and Algeria (Mehlman, 1989; Menard and Vallet, 1993) indicates that conceptions are also highly synchronised in free-ranging animals.

The physiological mechanisms underlying the high degree of reproductive seasonality in female Barbary macaques are not clear. Based on recent hormonal measures in females of the Gibraltar population, however, we provided the first evidence for seasonal changes at the ovarian level resulting in the restriction of ovulatory cycles to the mating season (Möhle *et al.*, 2005). In this respect, the Barbary macaque resembles other strictly seasonally breeding primates, including the rhesus and Japanese macaque, in which seasonal reproduction is also characterised by impaired ovarian function during the non-mating season (Nigi, 1975; Daily and Neill, 1981; Walker, Gordon and Wilson, 1983; Ziegler *et al.*, 2000).

Ovarian cycle characteristics

In contrast to many other macaque species (e.g. *M. tonkeana*: Thierry *et al.*, 1996; *M. silenus*: Shideler *et al.*, 1985, *M. nigra*: Thompson *et al.*, 1992; *M. fuscata*: Fujita *et al.*, 2001), little information exists on the characteristics of the female ovarian cycle and its endocrine correlates in the Barbary macaque. A major reason for this is the high probability of conception at the first ovulation of the mating season, making collection of data on cycle characteristics extremely difficult. However, in a recent study based on hormone analysis in faeces carried out on females of the Gibraltar Barbary macaque population, we have been able to describe the pattern of oestrogen and progestogen excretion associated with follicular development, ovulation, luteal function and early pregnancy (Möhle *et al.*, 2005). Figure 1 shows representative hormone profiles for one of our study females which experienced one non-conception cycle followed by conception in the second cycle. The endocrine profiles are characterised by a period of low concentrations of both oestrogens and progestogens at the beginning of the mating season in October/early November. Gradually rising oestrogen levels at the end of November/ early December followed by a sharp progestogen increase indicate follicular development and the onset of the luteal phase (post-ovulation), respectively. Progestogen levels then declined to baseline, which together with the occurrence of menstruation, indicated the absence of pregnancy. Following a second oestrogen rise and ovulation at the beginning of January, a sustained increase in progestogen levels accompanied by a marked elevation in oestrogens two to three weeks later indicated that conception had occurred. Composite profiles of oestrogen and progestogen excretion during five conception cycles and early pregnancies are shown in Figure 2 (panels b, c).

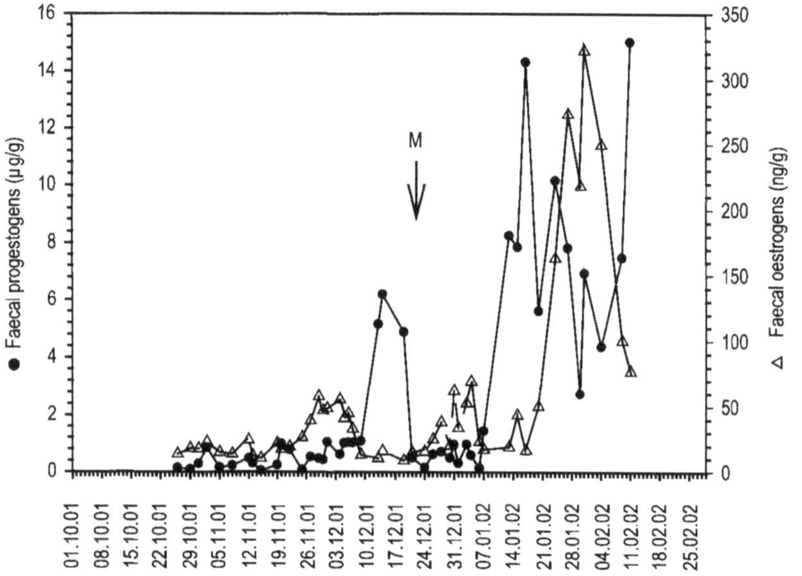

Figure 1. Faecal progestogen and oestrogen profiles during the breeding season in an individual female Barbary macaque. M = Menstruation. Figure adapted from Möhle *et al.* (2005)

Pooling results from non-conception cycles over a three year period, we have also been able to generate the first data on ovarian cycle length and its component phases. Using the defined progestogen rise and fall as a marker for the onset and end of the luteal phase of the cycle, a mean cycle length of 29.9 ± 2.9 days (mean ± SD; range 27-34 days, n=7), with a follicular phase length of 15.3 ± 1.6 days (range 14-18 days, n=4) and a luteal phase length of 15.6 ± 3.2 days (range 10-18 days, n=6) was found. Although based on relatively few cycles/females, the data indicate that in terms of overall cycle length and its component phases, the Barbary macaque is similar to most other species of its genus (e.g. Takahata, 1980; Wehrenberg *et al.*, 1980; Nieuwenhuisjen *et al.*, 1985; see also Table 4.1 in Bercovitch and Harvey, 2004), with the exception of the Sulawesi

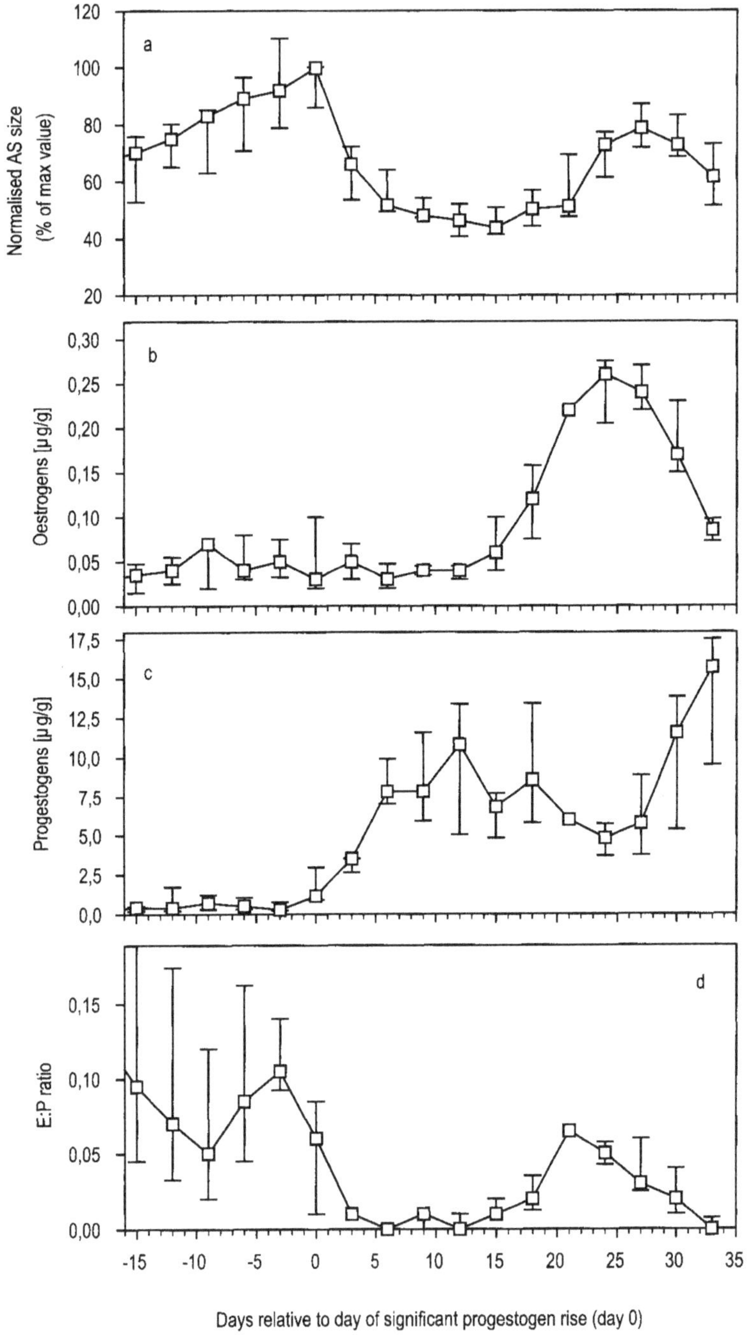

Figure 2. Composite profiles (median ± 25/75% percentiles) of faecal oestrogen (b) and progestogen (c) concentrations and the oestrogen to progestogen (E:P) ratio (d) in relation to normalised anogenital swelling (AS) size (a) during the conception cycle and early pregnancy (n=5 females). Data were plotted as 3-day blocks normalised to the day of the post-ovulatory progestogen rise. Figure adapted from Möhle *et al.* (2005)

macaques and the pig-tailed macaque which appear to have a markedly longer cycle length of about 35-40 days (Haddidian and Bernstein, 1979; Thomson *et al.*, 1992; Matsumura, 1993; Thierry *et al.*, 1996).

Data on gestation length in the Barbary macaque are also limited. Based on the interval between the end of the receptive period during the conception cycle and the date of birth, Küster and Paul (1984) reported a gestation length of 164.7 ± 6.1 days for 33 births observed at Affenberg Salem. Our own limited data, based on endocrine timing of ovulation in Gibraltar females (165.3 ± 3.7 days; n=6), are consistent with this. Thus, gestation length in the Barbary macaque falls within the range reported for other macaques (162.7 to 176.6 days; Bercovitch and Harvey, 2004).

Anogenital swelling (AS) characteristics

As in many other cercopithecoid primates living in multimale multifemale groups (Dixson, 1983; van Schaik *et al.*, 1999), the Barbary macaque shows a conspicuous swelling of the female anogenital region. Among macaques, this type of exaggerated swelling is a typical feature of species belonging to the *silenus-sylvanus* group, whereas swellings are greatly reduced in the members of the *fascicularis* group and are completely absent in the species belonging to the *sinica* and *arctoides* groups (Fooden, 1976; Dixson, 1983). In general, the variation in swelling patterns among the macaques can be explained by the degree of reproductive seasonality, with strictly seasonally breeding species (e.g. *M. radiata, M. sinica*) usually lacking exaggerated swellings and those with intermediate swellings (e.g. *M. fascicularis*) being more seasonal than those with fully exaggerated swellings (e.g. Sulawesi macaques). Thus, the Barbary macaque is unusual in being highly seasonal, yet having a very pronounced anogenital swelling.

In line with the pattern of ovarian activity, the occurrence of AS in the Barbary macaque is highly seasonal, being largely restricted to the mating season (Küster and Paul, 1984; Möhle *et al.*, 2005). Here, the onset of AS appears to be highly synchronised between females within a group (Möhle *et al.*, 2005) and is presumably related to closeness in timing of the onset of ovarian activity. Our recent studies have provided new information on quantitative changes of AS size in relation to known reproductive stages in a group of naturally reproducing females of the Gibraltar Barbary macaque population (Möhle *et al.*, 2005). Based on video imaging techniques to assess different size measures (e.g. width, height, depth), we have demonstrated that the swelling shows a well-defined cyclical pattern during the ovarian cycle, similar to that reported for other macaque species exhibiting exaggerated sex skin swelling (e.g. *M. tonkeana*: Thierry *et al.*, 1996; *M. nigra*: Thomson *et al.*, 1992; *M. fascicularis*: Engelhardt *et al.*, 2005).

More specifically, by combining the size measures of AS throughout the female cycle with faecal hormone measurements, Möhle *et al.* (2005) were able to show that, despite inter-individual variation in absolute AS size, the swelling is characterised by a progressive increase from the early to the late follicular phase in association with rising oestrogen (E) to progestogen (P) ratios followed by rapid detumescence after the onset of the luteal phase and declining E:P values.

Although the period of maximum AS always seems to encompass the time of ovulation, its timing within this period appears to be highly variable (Möhle *et al.*, 2005, M. Heistermann, unpublished). As in other species of macaques (e.g. Thierry *et al.*, 1996; Engelhardt *et al.*, 2005), changes in AS size in the Barbary macaque thus appear to provide information on the probability of ovulation (which males might use to help time their mating decisions) but not on its precise timing. Other functions in a non-reproductive (i.e. social) context should also be considered, particularly in view of its unusual occurrence in a species with a very restricted breeding season.

Reproductive behaviour

Reproductive behaviour in the Barbary macaque is more or less restricted to the short mating season from October to February (Taub, 1980; Küster and Paul, 1984). As for all other species of the genus *Macaca*, the Barbary macaque lives in multimale multifemale groups with a promiscuous mating system. However, high levels of apparently indiscriminate mating activity of females with multiple males of all age classes (Taub, 1980; Small, 1990) indicate a degree of promiscuity at the extreme end of the range for the genus (Small, 1990; Küster and Paul, 1992). Small (1990) proposed that this highly promiscuous mating pattern may be a direct consequence of the high degree of synchronisation of female ovarian cycles (see above) and a resulting low degree of defensibility of fertile females by males. Such a situation is in contrast to that seen in less seasonally reproducing macaque species, in which males are often able to monopolise and/or consort females during times of peak receptivity (e.g. *M. fascicularis*: Engelhardt *et al.*, 2006; *M. tonkeana*: Aujard *et al.*, 1998).

Although several studies have described the overall mating pattern in Barbary macaques, only limited information exists on whether timing of mating is indiscriminate in relation to female reproductive status or shows a distinct temporal pattern during the female's ovarian cycle as in other macaques (Dixson, 1998). Studies by Taub (1980) and, in particular, Küster and Paul (1984) have reported that, despite the highly promiscuous nature of the species, females exhibit distinct mating peaks which are separated by periods without any sexual activity. Using birth dates to estimate the approximate period of conception, the data from Küster and Paul´s study indicated a gradual increase in copulation rate throughout the 4 weeks preceding conception, with highest frequencies being seen during the presumed week of conception. Whilst this indicates that female reproductive status has an influence on the timing of mating behaviour, the degree to which copulatory activity in the Barbary macaque is related to ovarian cycle stage and female endocrine status was not addressed. Faecal hormone profiles generated in our recent study on females of the Gibraltar Barbary macaque population (Möhle *et al.*, 2005), have helped us to investigate this issue in more detail. Figure 3 shows copulation rates in relation to oestrogen and progestogen profiles throughout three ovarian cycles in one of our study females. As can be seen, the occurrence of copulations was mainly confined to the mid-to-late follicular phase of the cycle with a cessation of mating usually occurring around the time of the progestogen increase, signalling the onset of the luteal phase. Moreover, high levels of copulatory activity appear to be closely associated with elevated E:P concentration ratio. Pooled data from all study females (n=11; Figure 4) indicate that despite a high individual variation in copulation rates, particularly during the follicular and peri-ovulatory phases of the cycle, the overall mating pattern throughout the cycle was highly consistent between females. With respect to temporal changes in mating activity during the menstrual cycle and its underlying hormonal correlates, the Barbary macaque shows thus a pattern typical for the genus (e.g. *M. tonkeana*: Aujard *et al.*, 1998; *M. mulatta*: Wallen., 1984; *M. silenus*: Lindburg *et al.* 1985; *M. fuscata*: Fujita *et al.*, 2004; *M. arctoides*: Murray *et al.*, 1985) and other Old World primates (see Dixson, 1998).

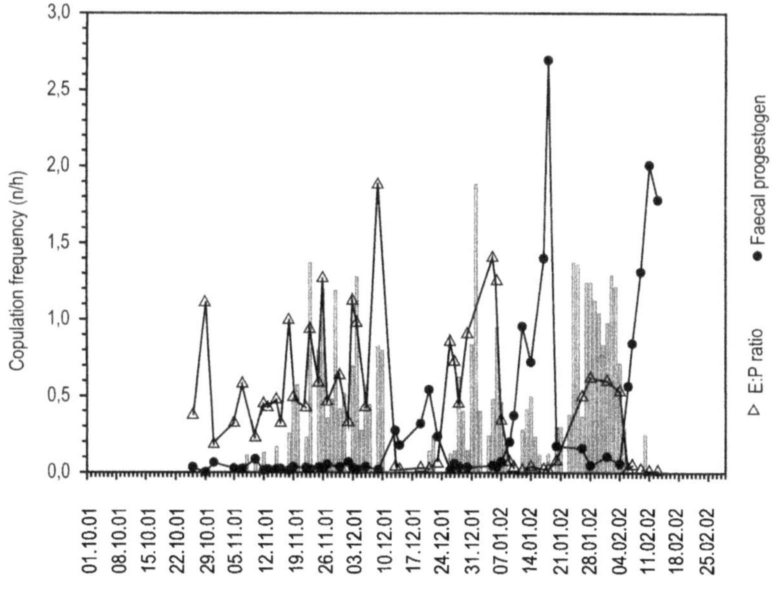

Figure 3. Distribution of copulations (grey bars) in relation to the profiles of faecal progestogens and the oestrogen to progestogen (E:P) ratio in an individual female Barbary macaque

In addition to mating behaviour, there is also some, although indirect, evidence that other sexual behaviours, including female proceptive behaviour, change during the female cycle (Taub, 1980; Small, 1990). The precise relationship of these behaviours with specific ovarian cycle stages and endocrine changes, however, awaits further investigation.

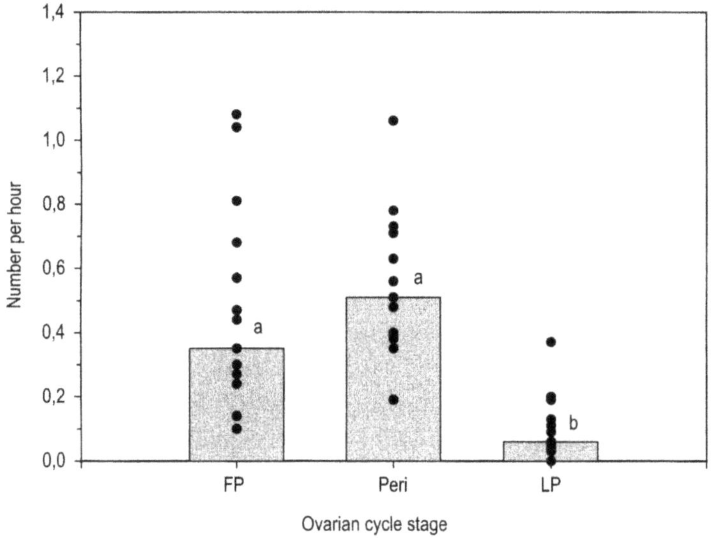

Figure 4. Copulation frequency in relation to ovarian cycle stage. Bars indicate median values calculated over 13 cycles (11 females). Circles represent individual cycles. FP = follicular phase, Peri = periovulatory phase, LP = early luteal phase as defined on the basis of the faecal progestogen profiles. Repeated Measures ANOVA: $p<0.001$; different superscripts indicate significant differences between stages (Student-Newman Keuls test, $p<0.001$)

Post-conception events

In addition to behavioural and morphological events occurring during the ovarian cycle, the Barbary macaque also shows a distinct period of mating behaviour during pregnancy (Küster and Paul, 1984; Small, 1990). This so-called post-conception oestrus (PCE) occurs 3-4 weeks after the onset of gestation and lasts for about 10-15 days (Küster and Paul, 1984). Although the presence of PCE has also been described for other macaque species, such as the stumptail macaque (Nieuwenhuijsen *et al.*, 1986), long-tailed macaque (van Noordwijk, 1985), rhesus monkey (Bielert *et al.*, 1976; Wilson *et al.*, 1982), and Japanese macaque (Takahata, 1980), species appear to differ considerably in terms of the frequency of PCE during pregnancy, its duration and timing relative to conception. For example, while Barbary macaque females show only one temporally discrete PCE period, up to four of these oestrus periods have been described for *Macaca fuscata* (Takahata, 1980). On the other hand, the PCE period in the stumptail macaque appears to be extended, with regular mating activity being observed during the entire first two months of gestation, but without a clear peak in copulation frequency during this period (Nieuwenhuijsen *et al.*, 1986). When the number of copulations during PCE is compared to that during the peri-ovulatory period, existing data for the Barbary macaque are equivocal. Küster and Paul (1984) reported that in most females copulation frequency and ejaculation rates were lower during PCE than during the receptive period of the conception cycle (although no empirical data are given). Small (1990), however, stated that copulation rates did not differ significantly between conception and post-conception cycles, emphasising that in half of her study females copulation rates were actually higher during PCE. Our own recent study on the Barbary macaques of the Middle Hill Group in Gibraltar found an overall lower frequency of both total and ejaculatory copulations during PCE compared to that during the peri-ovulatory phase of the ovarian cycle (M. Heistermann *et al.*, in preparation). There was, however, a high degree of variation among females, suggesting that individual characteristics, such as social status, age, familiarity to males etc. might influence the degree of mating activity during the post-conception cycle.

The period of post-conception mating in the Barbary macaque is also associated with a marked increase in anogenital sex skin swelling (Küster and Paul, 1984; Möhle *et al.*, 2005) which reaches on average 80% of the maximum size of that of the preceding fertile cycle (Möhle *et al.*, 2005; Figure 2a). Although the occurrence of anogenital swellings following conception has also been observed in other macaque species (e.g. *M. assamensis*, *M. nemestrina*, *M. fascicularis*, see compilation in Bercovitch and Harvey, 2004), these appear to be less conspicuous and/or regular than in the Barbary macaque. The only other species where a similar striking pattern of post-conception swelling has been reported is the mangabey (Gordon *et al.*, 1991), in which it occurs, however, at a later stage of gestation.

The physiological correlates underlying post-conceptional oestrus events in macaques are almost unknown. In our recent study on the Gibraltar Barbary macaques we have, however, clearly demonstrated that the occurrence of the post-conception swelling is closely associated with a marked increase in oestrogen concentrations and a pronounced rise in the oestrogen (E) to progestogen (P) ratio (Möhle *et al.*, 2005; Figure 2a, d). Thus, post-conception swellings in the Barbary macaque appear to be influenced by hormonal control mechanisms similar to those that affect swelling during ovulatory cycles, a finding also demonstrated for the mangabey (Gordon *et al.*, 1991). Since it is well-known that oestrogens enhance female attractivity and exert a stimulatory effect on female sexual behaviour while progesterone antagonises this effect, it is reasonable to assume that the occurrence of post-conceptional mating activity and swelling in the Barbary macaque are also stimulated by the change in the E:P ratio. Comparable data on this aspect are hardly available for other macaque species. In the rhesus monkey (Bielert *et al.*, 1976; Wilson *et al.*, 1982) and long-tailed macaque (Engelhardt *et al.*, in press), however, post-conception mating appears to be also associated with an elevated E:P ratio, suggesting a similar hormonal underpinning of post-conceptional events in macaques generally.

The presence of a clearly delimited period of anogenital swelling and mating activity during early gestation in the Barbary macaque raises intriguing questions concerning its functional significance. It has been proposed that mating during non-fertile periods in primates may be part of a female strategy helping to confuse paternity (and thereby reducing infanticide risks, Hrdy, 1974) or increasing male parental investment (Taub, 1980). However, to our knowledge there are no convincing empirical data demonstrating if and to what extent this may apply to the Barbary macaque, or indeed any other macaque species. In this respect, studies exploring the pattern of socio-sexual behaviours during post-conception cycles in comparison to that during ovulatory cycles could help to elucidate whether male Barbary macaques are able to distinguish between the two cycle types and, if so, how this influences the reproductive strategies of both sexes. This will in turn provide the basis for a better understanding of the function of the conspicuous swelling appearance and mating behaviour occurring at a time when fertilisation is no longer possible.

Acknowledgements

We are grateful to Dr. John Cortes, Gibraltar Ornithological and Natural History Society (GONHS) and Mark Pizarro, Gibraltar Veterinary Clinic, for their support and co-operation. We also thank Damian Holmes, Dale Laguea, Michael Whanon, Roger Rutherford and Paul Rocca for help with animal trapping, sample collection and transportation.

References

Aujard F, Heistermann M, Thierry B and Hodges JK (1998) Functional significance of behavioral, morphological, and endocrine correlates across the ovarian cycle in semifree ranging female Tonkean macaques. *American Journal of Primatology* **46** 285-309

Bercovitch FB and Harvey NC (2004) Reproductive life history. In *Macaque Societies: A Model for the Study of Social Organization* pp 61-83 Eds B Thierry, M Singh and W Kaumanns. Cambridge University Press, Cambridge

Bielert C, Czaja JA, Eisele S, Scheffler G, Robinson JA and Goy RW (1976) Mating in the rhesus monkey (*Macaca mulatta*) after conception and its relationship to oestradiol and progesterone levels throughout pregnancy *Journal of Reproduction and Fertility* **46** 179-187

Dailey RA and Neill JD (1981) Seasonal variation in reproductive hormones of rhesus monkeys: anovulatory and short luteal phase menstrual cycles *Biology of Reproduction* **25** 560-567

Dixson AF (1983) Observations of the evolution and behavioural significance of "sexual skin" in female primates *Advances in the Study of Behaviour* **13** 63-106

Dixson AF (1998) *Primate Sexuality* Oxford University Press, Oxford

Engelhardt A, Hodges JK, Niemitz C and Heistermann M (2005) Female sexual behavior, but not sex skin swelling, reliably indicates the timing of the fertile phase in wild long-tailed macaques (*Macaca fascicularis*) *Hormones and Behavior* **47** 195-204

Engelhardt A, Heistermann M, Hodges JK, Nürnberg P and Niemitz C (2006) Determinants of male reproductive success in wild long-tailed macaques (*Macaca fascicularis*)- male monopolisation, female mate choice or post-copulatory mechanisms? *Behavioural Ecology and Sociobiology* **59** 740-752

Engelhardt A, Hodges JK, and Heistermann M (2006) Post-conception mating in wild long-tailed macaques (Macaca fascicularis): characterization, endocrine correlates and functional significance Hormones and Behavior in press

Fooden J (1976) Provisional classification and key to living species of macaques (Primates: *Macaca*) *Folia Primatologica* **25** 225-236

Fujita S, Mitsunaga F, Sugiura H and Shimizu K (2001) Measurement of urinary and fecal steroid metabolites during the ovarian cycle in captive and wild Japanese macaques, *Macaca fuscata American Journal of Primatology* **53** 167-176

Fujita S, Sugiura H, Mitsunaga F and Shimizu K (2004) Hormone profiles and reproductive characteristics in wild female Japanese macaques (*Macaca fuscata*) *American Journal of Primatology* **64** 367-375

Gordon TP, Gust DA, Busse CD and Wilson ME (1991) Hormones and sexual behavior associated with postconception perineal swelling in the sooty mangabey (*Cercocebus torquatus atys*) *International Journal of Primatology* **12** 585-597

Hadidian J and Bernstein IS (1979) Female reproductive cycles and birth data from an Old World monkey colony *Primates* **20** 429-442

Hrdy SB (1974) Male-male competition and infanticide among the langurs (*Presbytis entellus*) of Abu, Rajasthan *Folia Primatologica* **22** 19-58

Küster J and Paul A (1984) Female reproductive characteristics in semi-free ranging Barbary macaques (*Macaca sylvanus* L. 1758) *Folia Primatologica* **43** 69-83

Küster J and Paul A (1992) Influence of male competition and female mate choice on male mating success in Barbary macaques (*Macaca sylvanus*) *Behaviour* **120** 192-217

Lindburg, DG (1971) The rhesus monkey in north India: an ecological and behavioral study. In *Primate Behavior: Development in Filed and Laboratory Research, Vol. 2* pp 1-106 Ed LA Rosenblum. Academic Press, New York

Lindburg DG Shideler SE and Fitch H (1985) Sexual behavior in relation to time of ovulation in the lion-tailed macaque. In *The Lion-tailed Macaque: Status and Conservation* pp 131-148 Ed P Heltne. Alan R. Liss, New York

Matsumura S (1993) Female reproductive cycles and the sexual behavior of moor macaques (*Macaca maurus*) in their natural habitat, South Sulawesi, Indonesia *Primates* **34** 99-103

Mehlman PT (1989) Comparative density, demography, and ranging behavior of Barbary macaques (*Macaca sylvanus*) in marginal and prime conifer habitats *International Journal of Primatology* **10** 269-292

Menard N and Vallet D (1993) Population dynamics of *Macaca sylvanus* in Algeria: an 8-year study *American Journal of Primatology* **30** 101-118

Möhle U, Heistermann M, Dittami J, Reinberg V and Hodges JK (2005) Patterns of anogenital swelling size and their endocrine correlates during ovulatory cycles and early pregnancy in free-ranging Barbary macaques (*Macaca sylvanus*) of Gibraltar *American Journal of Primatology* **66** 351-368

Murray RD, Bour ES and Smith EO (1985) Female menstrual cyclicity and sexual behavior in stumptail macaques (*Macaca arctoides*) *International Journal of Primatology* **6** 101-113

Nieuwenhuijsen K, Lammers AJJC, de Neef KJ and Slob AK (1985) Reproduction and social rank in female stumptail macaques (*Macaca arctoides*) *International Journal of Primatology* **6** 77-99

Nieuwenhuijsen K, de Neef KJ and Slob AK (1986) Sexual behaviour during ovarian cycles, pregnancy and lactation in group-living stumptail macaques (*Macaca arctoides*) *Human Reproduction* **1** 159-169

Nigi H (1975) Menstrual cycle and some other related aspects of Japanese monkeys (*Macaca fuscata*) *Primates* **17** 81-87

Shideler SE, Mitchell WR, Lindburg DG and Lasley BL (1985) Monitoring luteal function in the lion-tailed macaque (*Macaca silenus*) through urinary progesterone metabolite measurements *Zoo Biology* **4** 65-73

Small MF (1990) Promiscuity in Barbary macaques (*Macaca sylvanus*) *American Journal of Primatology* **20** 267-282

Takahata Y (1980) The reproductive biology of a free-ranging troop of Japanese monkeys *Primates* **21** 303-329

Taub DM (1980) Female choice and mating strategies among wild Barbary macaques (*Macaca sylvanus*). In *The Macaques: Studies in Ecology, Behavior and Evolution* pp 287-344 Ed DG Lindburg. Van Nostrand Reinhold, New York

Thierry B, Heistermann M, Aujard F and Hodges JK (1996) Long-term data on basic reproductive parameters and evaluation of endocrine, morphological, and behavioral measures for monitoring reproductive status in a group of semifree-ranging Tonkean macaques (*Macaca tonkeana*) *American Journal of Primatology* **39** 47-62

Thomson JA, Hess DL, Dahl KD, Iliff-Sizemore SA, Stouffer RL and Wolf DP (1992) The Sulawesi crested black macaque (*Macaca nigra*) menstrual cycle: changes in perineal tumescence and serum estradiol, progesterone, follicle-stimulating hormone, and luteinizing hormone levels *Biology of Reproduction* **46** 879-884

Van Noordwijk MA (1985) Sexual behaviour of Sumatran long-tailed macaques (*Macaca fascicularis*) *Zeitschrift für Tierpsychologie* **70** 277-296

Van Schaik CP, van Noordwijk M and Nunn CL (1999) Sex and social evolution in primates. In *Comparative Primate Socioecology* pp 204-231 Ed PC Lee. Cambridge University Press, Cambridge

Walker ML, Gordon TP and Wilson, ME (1983) Menstrual cycle characteristics of seasonally breeding rhesus monkeys *Biology of Reproduction* **29** 841-848

Wallen K, Winston LA, Gaventa S, Davis-DaSilva M and Collins DC (1984) Periovulatory changes in female sexual behavior and patterns of ovarian steroid secretion in group-living rhesus monkeys *Hormones and Behavior* **18** 431-450

Wehrenberg WB, Dyrenfurth I and Ferin M (1980) Endocrine characteristics of the menstrual cycle in the Assamese monkey (*Macaca assamensis*) *Biology of Reproduction* **23** 522-525

Wilson ME, Gordon TP and Collins DC (1982) Serum 17ß-estradiol and progesterone associated with mating behavior during early pregnancy in female rhesus monkeys *Hormones and Behavior* **16** 94-106

Ziegler T, Hodges JK, Winkler P and Heistermann M (2000) Hormonal correlates of reproductive seasonality in wild female hanuman langurs (*Presbytes entellus*) *American Journal of Primatology* **51** 119-134

Sexual swellings as specific mate recognition systems and the Barbary Monkey (*Macaca sylvanus*) as a proxy for the last common ancestor of macaques

JW Froehlich[1] and DJ Froehlich[2]
[1] *Division of Mammals, National Museum of Natural History, Smithsonian Institution, Washington, DC, 20560 USA*
[2] *Vertebrate Paleontology Laboratory, JJ Pickle Research Campus, University of Texas, Austin, TX, 78712 USA*

Introduction

As the flamboyant, but almost flightless male peacock was for Darwin, the grotesque, seemingly pathological sexual swellings of many female Old World monkeys have perplexed primatologists for decades. The venerable baboons of ancient Egypt (albeit usually male; e.g. Swidell, 2006) may have raised this spectre first for the founders of Greek zoology. More certainly, the proximity of the Barbary macaque, and its prominent use by Galen in the foundation of European medicine, must have maintained its presence throughout the history of Western biology. In fact, it was used as an example of sexual selection by Darwin (1876) himself, wherein males may select the most elaborate females as mates.

In recent years, much attention has been focused on the proximate reproductive and social functions of sexual swelling as an adaptive phenomenon within the sexual selection context (Dixson, 1998; Nunn, 1999; Stallmann and Froehlich, 2000). Following Hill's (e.g. 1974) meticulous literature review and comparative anatomical descriptions of the highly variable catamenial (i.e. monthly) changes among different species, however, there has been little systematic work. Almost no attention has been paid to speciation processes that ultimately derived the sexual swelling phenomenon, its remarkable modifications or its subsequent loss in different taxa, including our own lineage (but see Gangestad and Thornhill, 2004). Alternatively, it has been argued that sexual swellings (or very similar phenomena) may have evolved convergently on three (Nunn, 1999) or even five (Dixson, 1998) different occasions in Old World anthropoids, due to their spotty taxonomic distribution, and all presumably for the same adaptive functions related to sexual selection.

Recently, one of us has posited that the differentiation of new mate recognition systems (cf. Patterson, 1993), diverging during speciation, may better explain the origins of mostly qualitative differences (eg location and colour) of the phenomena among related species (Froehlich, 2003). This is not to deny that sexual selection of quantitative individual variations within each species does not play a major role in the subsequent elaboration of the various phenomena within each diverging lineage. After all, estrous characters will always comprise only a portion of the total specific mate recognition system (SMRS). As a portion operating only at limited times, these traits may be more flexible than pelage or other physical species differences when the species are totally isolated geographically.

Here we pursue this alternative account for the derivations of species novelties in anatomical phenomena, and the ultimate explanation for a presumably singular origin of the sexual swelling phenomenon in Old World anthropoids. Indeed, although not a prediction to be examined directly in this paper, we postulate that the predominantly red swellings may have evolved as estrous attraction signals to compensate for the loss of the vomeronasal organ (senses sexual pheromones). Enhanced red colour vision would then have coevolved with sexual swellings during Old World anthropoid origins, rather than as recently postulated for the discrimination of young leaves (Dominy, 2002; Dominy and Lucas, 2001; Lucas, *et al.*, 1998).

Consequently, the focus of this paper on the Barbary rock apes is largely methodological, using them as an important living link between Subsaharan primates and the marvellously diverse radiations of Asian macaques. Parenthetically, however, this study should also show how Barbary monkeys relate in their morphology to the hypothetical last common ancestor of macaques, or even of all members in the papionin tribe of monkeys. Rather than attending to variation and probable associated social function(s) within each species, or even closely related sister groups, we have sampled only one or two species of almost all genera of cercopithecine African monkeys that exhibit prominent visual estrous signals. Together with the Barbary monkey, we have also included only seven of the numerous species of Asian macaques as representatives of the extremes of variation that they represent.

We scored the colour, location, and relative expression of estrous signals among these taxa, and used cladistic analysis to portray this systematic variation optimally with respect to reasonable family trees of the study species. We predicted that a phylogenetic pattern approximating conventional classifications among closely related taxa using the conspicuous estrous characters should support the SMRS concept of speciation for the origin of these traits. If character displacements in these features follow other physical and molecular distinctions for each species, then these are presumed to have differentiated at the same time. In other words, we expect that the patterns of separate character enhancement or loss will simply follow the dividing and distinguishing of sister species and species groupings in the macaques, rather than reflecting any common patterns of behavioural ecology or ecogeography. A cladistic analysis uses the concept of parsimony (i.e. most simple solution preferred) to arrange taxa in a hierarchical tree that shows the pattern of shared changes in the character states with a minimal number of changes.

By this logic, we also anticipate that with a relatively constrained arsenal of possible variations in the estrous swelling phenomena, sister groups of species will be maximally differentiated from each other. They will display autapomorphies (i.e. unique novelties) or convergent homoplasies (i.e. traits not shared by close relatives but reoccurring more distantly) by the acquisition and loss of the same characters in new combinations. Concurrently, more distantly related species and genera may show repetitive patterns of convergence and parallelism in these same characters that will reduce or confuse the overall phylogenetic signal of the data. The net result of these predictions is the expectation that a phylogram (i.e. a cladogram or tree diagram with branch lengths proportional to the number of character changes or 'steps') showing patterns of relationship among taxa with estrous data will have long terminal branch lengths but very short basal lengths.

In line with our prediction that divergent patterns of estrous signal variation primarily serve the joint functions of intersexual recognition within species or species-lineages, and maximal distinction (i.e. reproductive isolation) from their geographical neighbours, the diverse estrous signals should show little consistent relationship to other reproductive variables, or to the different social structures in which they are exhibited. Within the geographically isolated species-lineages, such as the different species groups of macaques, the expected stabilising selection of mate recognition patterns (Froehlich, 2003) should also predict some continuity in the patterning of estrous displays. This prediction should contribute to moderately long branch lengths in a phylogram at these intermediate nodes.

Once again, let us emphasise that within each species, we would still expect quantitative elaborations of swelling size, colour intensity, or cycle duration to be correlated with reproductive variables among individuals, such as the debatable advertising of female quality with relative swelling size (Domb and Pagel, 2001). The sexual selection of these traits, however, may be secondary or supplemental to their differential origins within each species (Patterson, 1993). Therefore, we consider it unlikely that broad, intergeneric or subfamily comparative studies will yield a convincing, singular explanation of sexual selection for specific, qualitative characters (e.g. Nunn, 1999; Nunn, *et al.*, 2001; Zinner, *et al.*, 2004). This may partially explain the degree of controversy about the subject that has prevailed for the past few decades (e.g. Stallmann and Froehlich, 2000).

Methods

Judith Masters (1993) argued that mate recognition in the nonhuman primates has already occurred when cohesive, stable social groups are established, prior to any particular estrous cycle and the opportunity for sexual selection. Thus, we have constructed a character analysis that attempts to score all potentially utilised visual patterns of body shape, size, pelage pattern, and colour in both sexes. Additionally, we have borrowed data on reproductive cycling and social structure posted on the internet (Nunn, *et al.*, 2001). Eventually, in future research, these non-epigamic variables may enable us to test the prediction that they show little consistent relationship with specific estrous patterns. For the conspicuous catamenial changes used in this study, we have tried to score separately all aspects of perineal topology and the associated patterns of differential swelling and colouration in its various subparts.

The resulting character matrix has 84 traits (see Table 1) defined with the goal of scoring all of the potential variation in visual mate recognition systems, including estrous swellings, in all Old World anthropoid primates, including many taxa not included in this study. As a work-in-progress (the matrix is available on-line from either author) the present analysis utilised at most 69 parsimony-informative characters from this matrix; several characters were either monomorphic or unique so they provided no information about the taxa under study (parsimony attempts to determine the minimum number of character changes in a hypothetical phylogeny and the direction or polarity of these changes along each branch). For brevity, the table lists the characters by descriptive name, but omits the various character states and the scores for each species. The traits were either nominal (i.e. presence/absence) or ordinal (i.e. degree of expression as in never/seldom/common, or absent/narrow/wide). When these traits described colour they were often scored with multiple, partially nominal character states ("N" in the table). Nevertheless, they were often sequenced by shade (e.g. facial colour contrast was scored as none/dark/grey-brown/white) or hue (e.g. male perineal skin colour was scored as flesh/pink/red/purple/blue). When there was no apparent relationship to these colour sequences, the character was scored as simply nominative (e.g. estrous colour bright blue?).

Although we scored many characters in multiple states (i.e. ordered), rather than with binary presence/absence, we eventually decided to treat almost all of them as unordered (i.e. nominal) in the cladistic analysis presented here. At this point, only tail length was considered an ordinal, multiple state character, although its complete loss could still conceivably occur as a single step. For the size or intensity of other characters we were more interested in describing their degree of expression relative to adjacent features, rather than assuming an ordered sequence of evolutionary changes in any one feature. Cladistic analysis cannot adequately consider such covariance except in the combination of characters that change together along one tree branch.

The unordered definition of characters also seemed appropriate for graded patterns such as colour, since it was not clear that changes in state from blue to red, for example, would necessarily have purple as a transitional evolutionary stage or step. When some traits were polymorphic, they also did not always display a combination of adjacent, 'ordered' character states (i.e. a character might be scored as white or red, but not pink). Additionally, an earlier cladistic analysis that treated the data as ordered, while useful for pelage traits (see below), was unable to deal adequately with the catamenial characters. The primary effect of treating multiple state characters as unordered is that any change in the character is considered a single step, regardless of the actual number of 'graded changes' presumed in the ordered sequence. In other words, changes from large to absent or from large to small are both treated as single steps. The effect of this strategy on minimising steps is obvious, but there is a concomitant loss of information.

More importantly, an unordered character analysis of catamenial characters is also a conservative treatment of the data since there is usually a fair amount of variability (i.e. polymorphism) in expression among primate populations. Since some taxa were scored from only a few photographs, it is also reasonable to assume that these attempts to score the central tendencies of qualitative variability are only provisional approximations. Moreover, the best photographs or the most detailed descriptions (e.g. Hill, 1966; Rowell, 1977) are usually from captive animals, where nutritional or stress-mediated hormonal deviations may alter the central tendency of character expression in the wild. In fact, this disturbance has been manipulated in controlled research on the topic (e.g. Dixson, 1998). Age and status differences may also affect the appearance of the character in a limited record of photographs. Indeed, several taxa exhibit distinctive estrous traits in adolescence (Anderson and Bielert, 1994), and five of our characters score this variation, but it is not clear how well recognised this pattern is in all species in the wild (e.g. several species of Sulawesi macaque; Matsumara, 1993).

Finally, if we are correct in our prediction that estrous traits are significant components of SMRS among species, then their expression may vary in allopatry versus parapatry (i.e. living in isolated locations versus adjacent to each other), or between central and peripheral populations in the latter case. This is due to the expectation of greater character displacement for quick recognition when populations are in contact and mistakes possible (e.g. "The Sneetch Hypothesis" in Froehlich *et al.*, 1991). Thus, even careful observations and/or photographs in a wild population may vary from other populations of the same species.

Table 1. Character description for the matrix designed to score all Old World anthropoid visual features of potential mate recognition systems and associated reproductive variables. The 42 traits in the restricted analysis of catamenial changes in this paper are indicated with an X. Variables of three types are indicated by an O for multistate ordered characters, an N for two state nominal traits, and an "N" for multistate, approximtely sequenced nominal features. The character states of each variable indicate the total number of morphs used for scoring each variable.

No.	*Character Description*	*Selected subset*	*Variable type*	*Character states*
1	Adult Sex Ratio (M/F)		O	5
2	Single Male Groups?		N	2
3	Reproductive Group Size		O	5
4	Cycles to Conception		O	4
5	Pregnancy Matings?		N	2
6	Pregnancy Colour &/or Swelling		O	3
7	Interbirth Interval		O	4
8	Female Emigration?		O	3
9	Facial Skin Colour		O	4
10	Contrasting Eyelids?		N	2
11	Cheek Whiskers?		O	5
12	Facial Colour Contrast		O	4
13	Darker Crown Hair?		N	2
14	Male Mane or Mantel Hair		O	4
15	Facial Hairs Extend into a Ruff?		O	3
16	Beard?		O	3
17	Colour Nape & Shoulders		"N"	7
18	Darker Hips, Tail or Central Stripe?		N	2
19	Legs to Dorsum Contrast		O	3
20	Lighter Ventrum & Inner Legs?		O	3
21	Female Pelage Colour Contrast?		N	2
22	Chest Patch?		N	2
23	Male Coloured Rump Hair?		O	3
24	Male Perineal Skin Colour		"N"	5
25	Male Facial/Ventral Genital Mimicry?		N	2
26	Male Sex Skin Colour Change?		O	3
27	Male Genitalia Colour		"N"	7
28	Short, Erect Crown Hair?		O	3
29	Laterally Tufted Crown?		O	3
30	Whorled or Bared Crown?		O	4
31	Coronal Crest of Hair?		O	4
32	Forehead Balding?		O	4
33	Midparted Crown Hair?		O	3
34	Ear Tufts?		N	2
35	Maxillary Ridges or Crests?		O	3
36	Female Brow Ridges?		O	4

Table 1. Contd.

No.	*Character Description*	*Selected subset*	*Variable type*	*Character states*
37	Female Weight		O	7
38	Sexual Dimorphism (F/M weight)		O	5
39	Callosity Shape		"N"	6
40	Callosity Angle		O	3
41	Callosity Colour		"N"	5
42	Male Callosity Separation		O	3
43	Female Callosity Separation	X	O	3
44	Callosities Fully Surrounded by Integument	X	O	3
45	Gluteal Fields?	X	O	4
46	Rump Patch?	X	O	4
47	Tufted Tail?	X	O	3
48	Tail Curled?	X	N	2
49	Floppy Tail?	X	N	2
50	Tail Length	X	O	6
51	Anus Displaced to Tail Base	X	O	3
52	Swelling Primarily above Anus	X	O	3
53	Anus Puckered on Swelling	X	O	3
54	Anus Enveloped by Swelling	X	O	3
55	Vulval Margin Coloured &/or Swollen	X	O	5
56	Pendulous, Pseudoscrotal Lobes below Clitoris	X	O	3
57	Horizontal, Globular Pseudoscrotum over Pubes	X	"N"	3
58	Tail Base Swollen	X	O	3
59	Swelling Between Anus and Callosities	X	O	4
60	Perineal Swelling Extended Dorsolaterally	X	O	3
61	Sex Skin Surrounded by Vesicles?	X	N	2
62	Swelling Ventral and Lateral to Callosities	X	O	3
63	Swelling Surrounds Callosities (w/o merging)	X	O	3
64	Swollen Ventral Bags without Clitoral Lobe	X	O	3
65	Clitoral Lobe Size and Shape	X	"N"	3
66	Dorsal Swelling Extends into Pelage beyond Tail	X	O	3
67	Swelling above Anus shows Pseudotail Bulge?	X	N	2
68	Swollen Gluteal Fields?	X	N	2
69	Tail Subsumed by Swelling?	X	N	2
70	Tail Swollen into an oval Cylinder?		O	3
71	Tail Swollen into Rounded Ball?		N	2
72	Tail Swollen into a Pyramid w/ Broad Base?		N	2
73	Facial Colour Change?	X	N	2
74	Perineal Colour Change with Reduced Swelling	X	O	3
75	Estrous Colour	X	"N"	5
76	Estrous Colour Pastel Blue-grey?	X	N	2
77	Estrous Colour Bright Blue?	X	N	2

Table 1. Contd.

No.	*Character Description*	*Selected subset*	*Variable type*	*Character states*
78	Estrous Colour Translucent White?		N	2
79	Greater Subadult Swollen Tail Base?	X	N	2
80	Greater Subadult Swollen Clitoral Lobe?		N	2
81	Greater Subadult Pseudoscrotal Sack Swelling?	X	N	2
82	Greater Subadult Perineal or other Swelling?	X	N	2
83	Predominant Subadult Sex Skin Colouration?	X	N	2
84	Copius Mucous Secretions from Cervix?	X	N	2

Although the character analysis was designed as a work-in-progress to stimulate further sampling of as many taxa as possible, this preliminary investigation provided an initial test of the phylogenetic signal in the data and a model for the last common ancestor in this sequence of speciation events, at least for the macaques. Since good descriptions and illustrations were available for *Miopithecus talapoin*, but not Allen's swamp monkey, we selected this very small cercopithecin monkey as a potential outgroup to the tribe of interest in this study. Subsaharan papionins were represented by *Cercocebus albigena*, *Lophocebus atys*, *Mandrillus leucophaeus*, *Papio anubis*, *P. hamadryas*, and *Theropithecus gelada*. In addition to the Barbary monkey, we scored *Macaca fascicularis*, *M. mulatta*, and *M. radiata* as representatives of their various degrees of reduction or loss of estrous advertisements. As the closest relatives of the Barbary macaque, we included *M. nemestrina* and *M. silenus*, and two Sulawesi monkeys (*M. maurus* and *M. nigra*) as their derivatives.

We approached these data with a nested cladistic analysis, first grouping only the African taxa, including *M. sylvanus* to see its derivation of characters from its southern relatives. In this analysis, we used all 84 characters, albeit only 49 were parsimony-informative. Whereas our intention was then to include the Asian taxa within this constrained African cladogram, there was ambiguity in the precise relationship of *Macaca* to the various baboons. The analysis also failed to reflect current molecular-phylogenetic views on the diphyletic (i.e. not sharing a recent common ancestor) relationships of the two mangabey genera (Disotell, 1994; Harris and Disotell, 1998; Harris, 2000). Consequently, we decided to constrain the African tree further by defining *Cercocebus* as sister to *Mandrillus*, and *Lophocebus* as an unrestricted member of the baboon clade. In this second analysis, we included only the 42 characters (see the table) that described the topology of the perineal anatomy, including the tail, and the catamenial changes involving this anatomy. With this 'preferred' cladogram of eight taxa as an outgroup or constraint for the position of the macaque clade, we then included the remaining samples. In other words, this final cladistic analysis implicitly defined *Macaca sylvanus* as the basal macaque, relative to other papionins. This reduction of data to 42 characters was justified by the focus of this paper on estrous swellings and the relatively small number of Asian macaques included for sampling other epigamic and non-epigamic variation.

Results

Without specifying any constraints and including all 84 variables as unordered characters, Figure 1 displays the cladistic results for the eight African taxa. In this analysis, 15 characters were invariant and 20 were parsimony-uninformative as autapomorphies. As a preliminary test of our character definitions of all the epigamic and other features scored in this study as potential SMRS or associated characters, this analysis is remarkable in showing even slight information about the phylogeny or family tree of the study species. Although the cladogram was unrooted (i.e. not defining an ancestral taxon), *Miopithecus* consistently falls as the outgroup we expected. Baboons are also separated from mandrills, although this has only recently been recognised based on diphyletic relationships of the mangabeys in mitochondrial DNA patterns (Disotell, 1994).

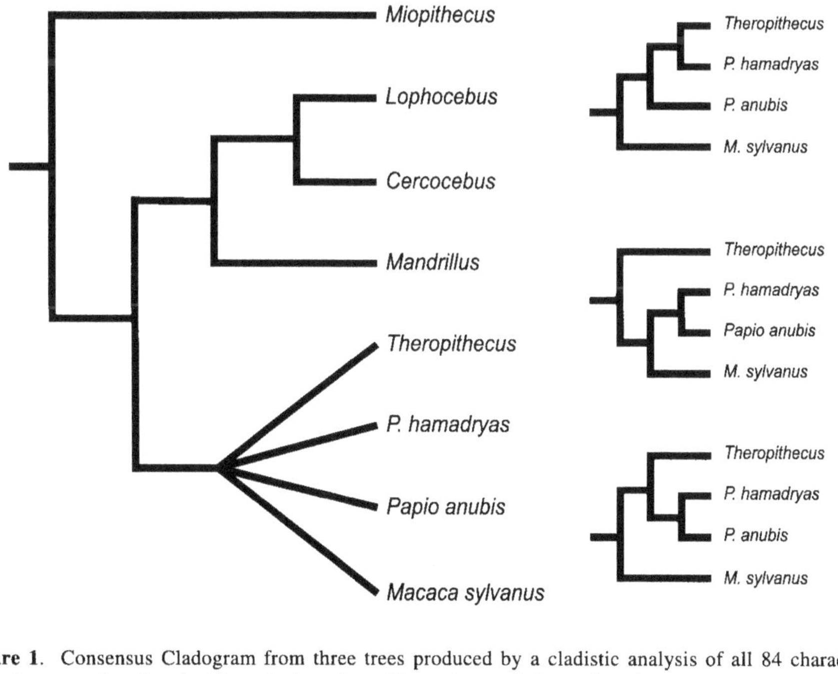

Figure 1. Consensus Cladogram from three trees produced by a cladistic analysis of all 84 characters treated as unordered and ultimately based on 49 parsimony-informative changes.

As anticipated among distant relatives, however, there are inconsistencies that may have more to do with convergence into similar ecological adaptations than actual family tree relationships. While *Cercocebus* is grouped with *Mandrillus*, as expected from recent studies that also included detailed morphology (Fleagle and McGraw, 2002), so too is *Lophocebus* as sister to the former. This again reflects the traditional view of monophyly of all mangabeys, and their somewhat similar ecological niches with mandrills. It supports some ecological adaptation for SMRS features. Also, while baboons are grouped together, as expected, their relationship to *Macaca sylvanus* and their own internal relationships are ambiguous in the three alternative, equally likely trees generated in this analysis, and this consensus tree is thus reduced to a polytomy for them. (A polytomy is a node with more than two branches, the historical relationships of which being unresolved.) Furthermore, within one additional step the entire consensus tree decayed to a great polytomy or 'rake', showing an overall lack of robusticity in the analysis.

Consequently, we constrained the diphyletic relationships of the mangabeys in subsequent analyses. For this study, we also reduced the characters to the 42 relating to perineal structures and their catamenial changes, in order to avoid any homoplasy (i.e. parallel evolution, convergence, or reversal) not relating to potential sexual selection during estrous. Since subsequent analyses would include macaques that retain long tail lengths, we also decided to treat this variable as ordered. Initially, we again included only the eight African taxa. In this analysis, 23 characters were parsimony-informative. Since this analysis produced only one tree, and it showed *Macaca sylvanus* basal to both *Lophocebus* and baboons (see that portion of Figure 2), we added the remaining macaque samples to this constrained tree. Thus, this constraint treated *M. sylvanus* as basal to all Asian members of the genus in a nested, stepwise procedure.

Figure 2 shows the Adams (i.e. relaxed) Consensus from seven, equally likely trees produced by the total analysis of all study species. This cladogram was based on 29 parsimony-informative characters. As displayed in the figure, there is another polytomy, this time with *M. radiata*, *M. nemestrina*, and two clades including Sulawesi monkeys and the *fascicularis* species group, respectively. The uncertainties of this figure are due to the ambiguous positions of the bonnet or pig-tailed monkeys as representing the basal member of Asian macaques, and whether the latter or the lion-tailed monkey are closest to the Sulawesi monkeys.

Two of the seven trees have *M. nemestrina* at the base of all Asian macaques. Clearly, this solution is minimising the homoplasy of convergence in the extreme sexual swellings seen in both pig-tailed monkeys and common baboons. The other five trees vary between *M. radiata* or *M. fascicularis* as the basal member or sister of the Asian macaques; they both suggest that the long tail lengths of these taxa may reflect the ancestral condition of the Asian clade. The *M. fascicularis* alternative more closely reflects mitochondrial and y-chromosome molecular phylogenies (Morales and Melnick, 1998; Tosi, *et al.*, 2000).

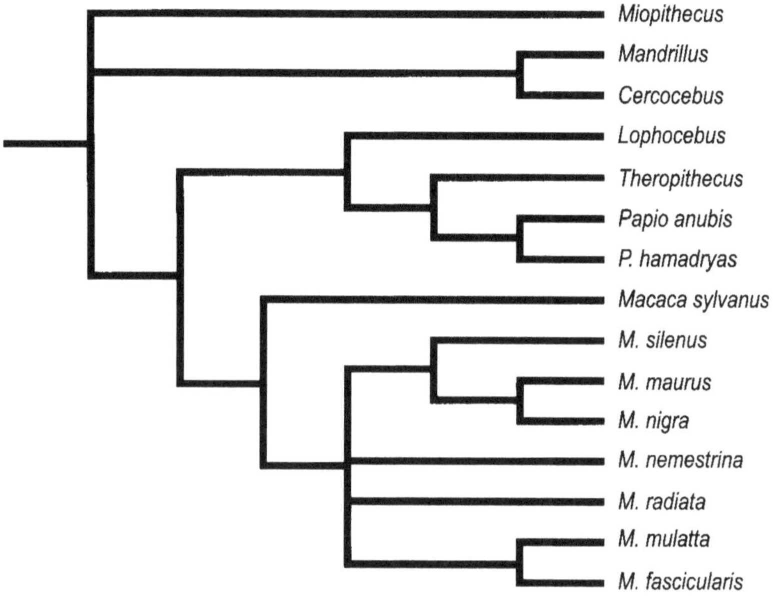

Figure 2. Adams Consensus Cladogram from seven trees produced by a cladistic analysis of 42 estrous characters treated as unordered (except for tail length) and constrained for the diphyletic positions of *Cercocebus* and *Lophocebus*. These relationships were first determined in a nested analysis of the upper eight taxa before the lower seven were added with this first (and single) tree as a constraint. In the first tree, 23 characters were parsimony-informative; in the final there were 29.

In two trees, the sister relationship to Sulawesi is also unresolved between *M. silenus* and *M. nemestrina*; this is also the case with some molecular data, although geography would favour the latter. Thus, the phylogram in Figure 3, as our preferred tree based both on internal analysis and the views of the most recent and complete mitochondrial DNA data (Evans *et al.*, 2003), is the one with a *M. nemestrina* and Sulawesi clade. This preference of trees is also supported by the combined basal branch lengths being almost twice as long as in the *M. silenus* alternative. In fact, the only trait at the base of the entire clade when *M. nemestrina* takes this position — swellings extended dorsolaterally — is also shared on the same branch when *M. silenus* is at the base.

Discussion

Without any constraints in the cladistic analysis, the epigamic characters are somewhat comparable to other morphological studies of the past (e.g. Delson and Dean, 1993). Molecular phylogenies have revolutionised systematic studies, allowing control of deep branching sequences so that patterns of parallel evolution and convergence can be identified and studied from an adaptive perspective. In the present study, the initial analysis (Figure 1) clustered *Lophocebus* with *Cercocebus*, but they also formed a clade with *Mandrillus*. Interestingly, this arrangement is midway between the former, morphological view of mangabey monophyly and the current molecular resolution that divides them between the mandrills and the baboons (Fleagle and McGraw, 2002).

In terms of the primary ecological distinction between the two mangabeys, it appears that *Lophocebus* is more arboreal while *Cercocebus* is semiterrestrial (Disotell, 1994). With respect to the epigamic features, we speculate that the spatial separations of individual members of *Cercocebus* and *Mandrillus* groups are somewhat similar and subject to the same spotty lighting and line-of-sight obstructions in the under-storey and forest floor. Under these conditions, visible estrous signals might be amplified for distance with respect to those monkeys higher in the canopy where social groups in single trees may be more cohesive. Alternatively, these signals might also be more diffuse (perhaps patchy) or muted to avoid being overly conspicuous to predators like hunting dogs, leopards, raptors, or people. (Canids and felids have colour vision superficially like human red/green deuteronopes, but apparently 'see' red with greater contrast as a 'purple'; raptors see red and even ultraviolet and could be attracted to the young near cycling females.).

For the 42 estrous characters displayed in the phylogram (Figure 3), *Cercocebus* and *Mandrillus* share distinctive pelage rump patches, bared integument surrounding the ischial callosities (i.e. 'sitting pads'), a bright pink or red anus distended from a surrounding, differently coloured swelling, and greatly amplified vulval margins. Interestingly, the latter character state is shared with *Miopithecus*, and possibly also in the common ancestral macaque, but it is either amplified or reduced in the various later macaque and baboon lineages. There are also homoplasies with these data that complicate interpretation, such as the rump patches in *Papio hamadryas* males, and in some Asian macaques of both sexes.

Nevertheless, the analysis additionally suggests pastel pink colouration, sometimes combined with blue tints, as the primitive sexual swelling condition, and it is informative to see this also retained in the Barbary monkey. However, this trait is homoplastically amplified to bright pink or red in *Mandrillus* and this intensification also seems to be the general condition in *Lophocebus* and baboons, and in all later macaques. The mandrills and Barbary monkey also share swellings extending into the pelage of the lower back, and again this occurs in some Asian macaques.

Although confounded by not unexpected homoplasy, the analysis of these data suggests that papionin estrous swellings may have initially evolved in the patchy sunlight and spatial separation of early semiterrestrial rainforest monkey populations. While attractive to other monkeys in the range of their newly evolving red colour sensitivity, these estrous patterns are also muted, patchy, and mutually contrasting (i.e. 'spotted') to reduce concomitant attraction of predators. In contrast, the more arboreal *Lophocebus* and its open country terrestrial relatives could concentrate on amplifying a single colour modality from bright pink to red, or even purple. Certainly, it is interesting that the Barbary macaque appears to have retained some of these primitive conditions in its habitat beneath the cedars of North Africa. It is also important to note that these reconstructions are consistent with the emphasis placed on ecological adaptation in the evolution of SMRS traits by the authors of this species concept (cf. Paterson, 1993; Masters, 1993).

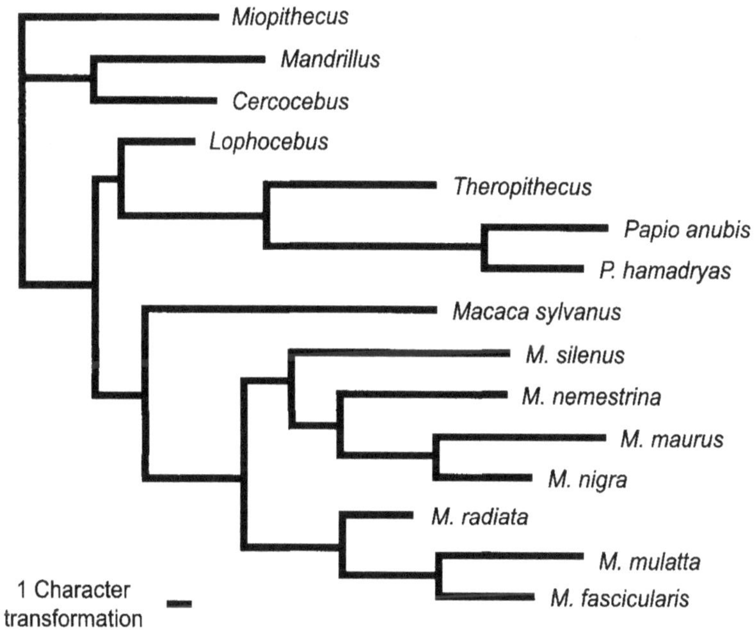

Figure 3. Phylogram based on the preferred of the seven trees produced in the analysis illustrated in Figure 2.

The very ambiguity in these basal data due to homoplasy is also consistent with our mate recognition hypothesis. Indeed, if estrous swellings had evolved as a singular, homogeneous pattern, subject only to runaway sexual selection in the study taxa, and with divergent speciation unassociated with the presence or size of the swellings, there should be more phylogenetic pattern in the basal relationships based on the expression of these perineal characters. That is, if these traits only responded to the same ecological selection pressures as other physical features, they should not be maximally distinguishable between sister species within the closely similar genera of the papionins. Outside of the papionins, it is this distribution pattern that has also led to the presumption of multiple origins for the estrous swelling trait.

Remarkably, when the analysis was artificially constrained by the known mangabey relationships, the position of *Macaca sylvanus* no longer produced a polytomy (cf. Figure 3 and Figure 2). Although living macaques, based strictly on estrous characters, do not fall in line with the suspected primitive relationships to *Cercocebus* and *Mandrillus*, the hypothetical last common ancestral macaque may have. Also, the actual placement of macaques in the constrained cladogram (Figure 2) is also not necessarily expected from molecular studies (cf. Disotell, 1996), in which macaques are usually basal to all other papionins. Nevertheless, for the purposes of studying relationships within the genus based on estrous traits, and especially for generating their hypothetical last common ancestor, we elected to keep macaques at the basal node in the baboon clade. Similarly, we also constrained the Barbary monkey position in the full analysis as basal to all other macaques.

Using the preferred phylogram and the steps in character change that it displays, the last common ancestor of macaques was very much like the Barbary monkey, but with an expected tail length roughly two-thirds of body length. This analysis also suggests that the base of this tail was probably swollen and/or surrounded by the sexual swelling. This may account for the subsequent loss of the tail in today's Barbary monkey, as posited for the equatorial Sulawesi macaques (Froehlich, 2003). For the mostly arboreal Sulawesi monkeys living today near sea level on the equator, it seemed unlikely that cold intolerance (i.e. frostbite) could account for extremely short tail lengths; rather, the lack of hygiene during monthly cycling may select for shorter tail lengths.

The last common ancestor of macaques most likely had ischial callosities that were fully separated in females, and the pastel pink sexual swellings extended up to the base of the tail, moderately enveloping the anus, and spreading laterally above the ischial callosities and dorsally into the lower back. Swellings did not, however, surround the ischial callosities even though the skin here was likely bare. The margins of the vulva were probably inflated but not exaggeratedly so, and the labia majora extended only moderately below a modest clitoral lobe, unlike the greatly amplified condition seen in *M. sylvanus*, or in *M. nemestrina* and *Papio*. Finally, the pastel coloration retained in the Barbary macaque probably represents the ancestral condition, shared with the *Cercocebus* clade, with bright red then being intensified in parallel in the baboons and later macaques. Given the estrous swelling homoplasy already discussed in these basal relationships, this hypothetical ancestral macaque cannot be further refined, nor can a hypothetical ancestor for the other papionins. Nevertheless, the hypothetical macaque's catamenial traits are in many respects fairly similar to the Barbary monkey.

The same conclusion can be drawn from the pelage data, based on an unconstrained analysis (not shown here) that stepped *Lophocebus*, *Cercocebus*, and *Mandrillus* sequentially into the cladogram. Presumably due to convergence in social structure, *Theropithecus* was also paired with the latter. Sister to the remaining baboons, the Barbary monkey was then placed at the base of all macaques. (In this ordered analysis of the entire data set, the Barbary macaque was not constrained to the base of its genus; 69 characters were parsimony informative.) This analysis predicts that the last common macaque ancestor had ischial callosities that were triangular or bar-shaped, bright pink, arranged at a moderate angle to each other, and meeting in the midline in males, but not fusing. Given the relatively high latitude and mountain habitats of *M. sylvanus*, body size and its effect on dimorphism are difficult to predict, but the data suggest a fairly large animal, larger than most Asian macaques and with minimal sexual dimorphism; females probably also had fairly prominent brow ridges. The short, erect crown hair of the Barbary monkey is probably unique as the ancestor may have had a midparted hairdo similar to *M. silenus* and *M. nemestrina*. The males probably did not have a distinctive rump patch, their perineums were slate or pink, and the genitalia were likely red in colour. The facial skin was probably grey or brown, darker than the present rock apes, but they also had contrasting eyelids, modest cheek whiskers with slightly lighter contrast, darker crown hair, a facial ruff, and only a slight beard. In most respects this description, as with the female catamenial characters, approximates the Barbary rock apes. This underscores their importance for any comparative or intensive studies of behavioural function in all of these epigamic traits, both pelage and sexual swellings.

If the basal relationships of all papionins are ambiguous, this does not appear to be the case with the distal twigs of the phylogram in Figure 3. As predicted, these appear to be robust, with branch length increasing from left to right in the tree. Despite the inconsistencies of different or similar social structures, the two baboon species are clustered and separated from geladas by long branch lengths. *M. sylvanus* is also quite distinctive, with many autapomorphies that partially belie the previous conclusion of its relatively 'primitive' condition. While basal relationships have intermediate to small branch lengths in the *silenus* and *radiata* clades of Asian macaques, each individual branch is long. The Sulawesi and *fascicularis* clades are also well defined by intermediate branch lengths comprising several synapomorphic (i.e. shared, derived) traits.

Overall, then, this study provides some support for the SMRS hypothesis in the origin and evolution of variability in the estrous swelling phenomena of Old World cercopithecine monkeys. With a limited repertoire of potential qualitative variants, due to a relatively small region of the body, there is also a remarkable degree of homoplasy. This is especially the case in the surviving members of the older, basal branches of the papionins as they diverged from other cercopithecine and colobine monkeys, sometime around 9 to 14 million years ago (van der Kuyl, *et al.*, 2000). Even the position of macaques relative to other papionins is unclear with these data, discordant with most molecular data that show it to be an outgroup to the Subsaharan clades (Disotell, 2000).

Nevertheless, even at these basal nodes, the epigamic traits still show some phylogenetic signal, as with the consistent separation of *Miopithecus*. This is all the more noteworthy when the talapoin monkey has many of its own autapomorphies that parallel various traits in the baboons and macaques. For example, the diminutive *Miopithecus* has swellings completely surrounding the ischial callosities and a swollen tail base, as occurs in some macaques and baboons, but not in their basal branches. This commonality of many character states, even in parallel, implies the origin of a single biological system in the estrous swellings of all Old World anthropoids. Multiple origins would imply an impossible number of independent convergences, rather than just a single convergence in the sexual swelling phenomenon. Given the homoplasy occurring elsewhere in this study, these convergences in a very small monkey (sometimes called a dwarf guenon), that is clearly outside the papionins (Tosi, *et al.*, 2002), support a single origin of sexual swelling in the Old World monkeys, at least. That is, all examined species are in some sense 'derived' recombinations from the common ancestral form. However, without a larger sample of study species, especially the swamp monkey, we cannot attempt to reconstruct a last common ancestor at this time.

More importantly, these analyses show considerable phylogenetic signatures for the estrous characters in the terminal branches, demonstrating each species or species group has a distinctive combination of elaboration or reduction in the same characters reoccurring homoplastically in different combinations in other clades. It would seem clear that these traits differentiated in conjunction with the speciation process. They can then be stabilised in allopatric lineages or species groups, but they appear maximally differentiated from sympatric congenerics. Quite possibly, the extreme reduction or complete absence of sexual swellings in some Asian macaques has little to do with shared behavioural changes in their reproductive biology; but instead, they may represent just one more way to define SMRS distinctive from the neighbourhood of parapatric species. As these relationships based on the evolution of estrous characters are further refined by cladistic analyses, they should allow more precise examination of the social implications and adaptive functions (i.e. sexual selection) of each, just as molecular phylogenies have allowed us to test new models of morphological evolution.

In conclusion, we think it is worth repetition that the rock apes of Gibraltar and the Barbary Coast are a critical species in the evolution of Old World monkeys. They still possess many of the ancestral pelage and catamenial traits predicted in a common ancestral macaque, and even in an ancestral anthropoid. Sexual swellings that are pastel in colour but emphasising at least rudimentary red colour vision, extending along the midline from the base of the tail to a moderately swollen clitoris, and bordered by swollen, baggy labia that mimic scrota were in all likelihood part of the suite of characters that occurred for mate recognition and sexual selection at the outset of Old World monkeys. If the phenomenon of sexual swelling and its many manifestations are to be fully understood, then further studies on Gibraltar and in Morocco and Algeria are essential.

Finally, an evolutionary scenario, akin to Rudyard Kipling's *Just So Stories*, may clarify and simplify the relationship we see between runaway sexual selection and stabilising mate recognition for the same set of characters. With the first Old World anthropoid primate, compensating for a reduction of olfactory estrous signals, sexual swellings and red colour vision coevolved with sexual selection leading to rapid fixation of each, albeit with pastel hues as would also be expected in the case of discrimination of young leaves. Subsequently, sexual selection might act on variants of size and colour intensity, but novelties as in placement or incorporation of different anatomical structures would be resisted by the stabilisation of the SMRS within the species. With each new speciation, however, the intraspecific variability produced by sexual selection would be translated into interspecific distinctions, along with any novelties that might be fixed in small peripheral subpopulations. Subsequently, with each species divergence, two contrasting and novel sets of estrous attraction characters would be fixed by the SMRS process, followed by sexual selection amplifying the intensity of each. As Patterson (1993, p77) argued, characters like sexual swellings may have "evolved first as part of the SMRS and then evolved secondarily, to function in intermale conflict…"

Acknowledgements

A large number of people are responsible for the photographs that have been consulted in this study, and it is not practical to name all of them, or to cite all of the publications in which they occur, but we are immensely grateful for all of them. In a phenomenon that was previously considered monotypic, these records of variation, both systematic and individual, are crucial, and we hasten to thank future contributors as well. Nevertheless, we especially wish to thank Irwin Bernstein, Ben Evans, Lisa Jones-Engel, Don Lindburg, Ulricke Möhle, Michael Schillaci, Osamu Takenaka, and Bernard Thierry.

We have also benefited from much discussion and argument with colleagues and students over the past five years, even if some nonbiological anthropology graduate students thought the topic prurient and unacceptable for presentation to undergraduate students. We are particularly grateful for past criticisms and arguments about our ideas from Judith Masters.

References

Anderson CM and Bielert CF (1994) Adolescent exaggeration in female catarrhine primates *Primates* **35** 283-300

Darwin C (1876) Sexual selection as related to monkeys *Nature* **15** 18-19

Delson E and Dean D (1993) Are *Papio baringensis* R. Leakey, 1969, and *P. quadratirostris* Iwamoto, 1982, species of *Papio* and *Theropithecus*? In Theropithecus*: the Rise and Fall of a Primate Genus* pp 125-156 Ed NE Jablonski. Cambridge University Press, Cambridge

Disotell TR (1994) Generic level relationships of the Papionini (Cercopithecoidea) *American Journal of Physical Anthropology* **94** 47-57

Disotell TR (1996) The Phylogeny of Old World monkeys *Evolutionary Anthropology* **5** 18-24

Disotell TR (2000) Molecular systematics of the Cercopithecidae. In *Old World Monkeys* pp 29-56 Eds PF Whitehead and CJ Jolly. Cambridge University Press, Cambridge

Dixson AF (1998) *Primate Sexuality: Comparative studies of the Prosimians, Monkeys, Apes, and Human Beings* Oxford University Press, Oxford

Domb LG and Pagel M (2001) Sexual swellings advertise female quality in wild baboons *Nature* **410** 204-206

Dominy NJ (2002) Incidence of red leaves in the rainforest of Kibale National Park, Uganda: Shade-tolerators and light-demanders compared *African Journal of Ecology* **40** 94-96

Dominy NJ and Lucas PW (2001) Ecological importance of trichromatic vision to primates *Nature* **410** 363-366

Evans BJ, Supriatna J, Andayani N and Melnick DJ (2003) Diversification of Sulawesi macaque monkeys: Decoupled evolution of mitochondrial DNA *Evolution* **57** 1931-1946

Fleagle JG and McGraw WS (2002) Skeletal and dental morphology of African papionins: Unmasking a cryptic clade *Journal of Human Evolution* **42** 267-292

Froehlich JW (2003) Testing some theoretical expectations of sexual selection versus the recognition species concept in the speciose macaques of Sulawesi. In *Sexual Selection and Reproductive Competition in Primates: New Perspectives and Directions* pp 538-591 Ed CB Jones. American Society of Primatologists, Norman, Oklahoma

Froehlich JW, Supriatna J and Froehlich PH (1991) Morphometric analyses of *Ateles*: systematic and biogeographical implications *American Journal of Primatology* **25** 1-22
Gangestad SW and Thornhill R (2004) Female multiple mating and genetic benefits in human investigations of design. In *Sexual Selection in Primates: New and Comparative Perspectives* pp 90-113 Eds PM Kappeler and CP van Schaik. Cambridge University Press, Cambridge
Harris EE (2000) Molecular systematics of the Old World monkey tribe Papionini: Analysis of the total available genetic sequences *Journal of Human Evolution* **58** 235-256
Harris EE and Disotell TR (1998) Nuclear gene trees and the phylogenetic relationships of the mangabeys (Primates: Papionini) *Molecular Biology and Evolution* **15** 892-900
Hill WCO (1966) *Primates: Comparative Anatomy and Taxonomy. VI Catarrhini, Cercopithecoidea, Cercopithecinae* Interscience Publishers, Inc., New York
Hill WCO (1974) *Primates: Comparative Anatomy and Taxonomy. VII Cynopithecinae:* Cercocebus, Macaca, Cynopithecus Edinburgh University Press, Edinburgh
Lucas PW, Darvell BW, Lee PKB, Yuen TDB and Choong ME (1998) Colour cues for leaf food selection by long-tailed macaques (*Macaca fascicularis*) with a suggestion for the evolution of trichromatic colour vision *Folia primatologica* **69** 139-152
Masters JC (1993) Primates and paradigms: Problems with the identification of genetic species. In *Species, Species Concepts, and Primate Evolution* pp 43-64 eds WH Kimbel and LB Martin. Plenum Press, New York
Masters JC (1998) Speciation in the lesser galagos *Folia Primatologica* **69** (suppl 1) 357-370
Matsumura S (1993) Female reproductive cycles and the sexual behavior of moor macaques (*Macaca maurus*) in their natural habitat, South Sulawesi, Indonesia *Primates* **34** 99-103
Morales JC and Melnick DJ (1998) Phylogenetic relationships of the macaques (Cercopithecidae: *Macaca*), as revealed by high resolution restriction site mapping of mitochondrial ribosomal genes *Journal of Human Evolution* **34** 1-23
Nunn CL (1998) The Evolution of exaggerated sexual swellings in primates and the graded-signal hypothesis *Animal Behaviour* **58** 229-246
Nunn CL, van Schaik CP and Zinner D (2001) Do exaggerated sexual swellings function in female mating competition in primates? A Comparative test of the reliable indicator hypothesis *Behavioural Ecology* **12** 646-654
Patterson HEH (1993) *Evolution and The Recognition Concept of Species: Collected Writings.* Ed SF McEvey. The Johns Hopkins University Press, Baltimore

Rowell TE (1977) Reproductive cycles of the talapoin monkey (*Miopithecus talapoin*) *Folia Primatologica* **28** 188-202

Stallmann RR and Froehlich JW (2000) Primate sexual swellings as coevolved signal systems *Primates* **41** 1-16

Swedell L (2006) *Strategies of Sex and Survival in Hamadryas Baboons: Through a Female Lens* Pearson Education, Inc., Upper Saddle River, New Jersey

Tosi AJ, Morales JC and Melnick DJ (2000) Comparison of Y-chromosome and mtDNA phylogenies leads to unique inferences of macaque evolutionary history *Molecular Phylogenetics and Evolution* **17** 133-144

Tosi AJ, Buzzard P, Morales JC and Melnick DJ (2002) Y-chromosome data and tribal affiliations of *Allenopithecus* and *Miopithecus. International Journal of Primatology* **23** 1287-1299

van der Kuyl AC, Dekker JT and Goudsmit J (2000) Primate genus *Miopithecus*: Evidence for the existence of species and subspecies of dwarf guenons based on cellular and endogenous viral sequences *Molecular Phylogenetics and Evolution* **14** 403-413

Zinner DP, van Schaik CP, Nunn CL and PM Keppeler (2004) Sexual selection and exaggerated sexual swellings of female primates. In *Sexual Selection in Primates: New and Comparative Perspectives* pp 71-89 Eds PM Kappeler and CP van Schaik. Cambridge University Press, Cambridge

Philopatry of female macaques in relation to group dynamics and the distribution of genes - the case of the Barbary macaque (*Macaca sylvanus*)

N Ménard[1], M Lathuillière[1,2], E Petit[1], D Vallet[1], B Crouau-Roy[2]
[1]*UMR 6552, Ethologie-Evolution-Ecologie, Station Biologique, 35380 Paimpont, France*
[2]*UMR 5174 «Evolution et Diversité Biologique» EDB, Université P. Sabatier, 31000 Toulouse, France*

Introduction

Despite great species diversity, macaques show an apparent homogeneity in their social organisation, characterised by a multimale-multifemale group structure. Maximum group size is around 70-90 individuals except in provisioned conditions or rural areas, where groups can reach up to 140 individuals. The smallest groups include around 10 individuals, although there can be considerable variation depending on group life history or ecological conditions (for a review, see Ménard, 2004). There are also great intraspecific variations depending on ecological conditions. All macaque species are characterised by an extreme dispersal asymmetry of sexes. Whatever the species, males leave their natal group around sexual maturity while females are philopatric and form several clusters of related individuals within groups, known as matrilines. Dispersal patterns shape the demographic and genetic structure of groups and these may differ between species or even between populations within species. In addition, the characteristics of matrilines and their dynamics should be taken into account in studies of the distribution of genes and relatedness. In this chapter, we will review the state of our knowledge about the relationships between group dynamics and the distribution of genes with special reference to the influence of group fission. The paper will review information available for macaques in general, highlighting the particular case of the Barbary macaque.

How do dynamic events affect demographic and genetic structure in macaques?

Male dispersal

Since migrations are discrete events separated over time (and thus requiring long-term studies on several focal groups), relatively little is known about the dispersal patterns of male macaques. Nevertheless, some general trends can be drawn. It is generally recognised that in most macaque species, almost all males leave their natal group before reaching adulthood, that is around 3-6 years of age (*Macaca nemestrina*, Oi, 1990; *M. fuscata*, Sugiyama, 1976; *M. mulatta*, Melnick *et al.*, 1984), and in doing so avoid reproducing with their female relatives. Van Noordwijk and van Schaik (1985) reported for long-tailed macaques that no males were found in their natal groups by an estimated age of seven. In other species (e.g. *M. maurus)* however, male dispersal occurs later (7-9 years) and even mature males as old as 9 have been recorded in their natal groups (Okamoto, Matsumura and Watanabe, 2000). Various modalities of migration have been described according to the study or the species. Migrating males can experiment with a solitary life for several months (*M. fuscata*, Mori and Watanabe, 2003), join all-male bands (*M. thibetana,* Zhao, 1994) or rapidly integrate into a new group, avoiding a prolonged period of solitude and reducing the likelihood of predation (*M. nemestrina*, Oi, 1990; *M. fascicularis*, van Noordwijk and van Schaik, 1985). Most males transfer into neighbouring groups (about 82% in wild *M. mulatta*, Melnick *et al.*, 1984; *M. fascicularis*, van Noordwijk and van Schaik, 1985) although some may migrate as far as 60km (*M. fuscata*, Yoshimi and Takasaki, 2003).

Matriline characteristics and group fission

In contrast to males, females stay in their natal group all their life. Female philopatry results in the existence of clusters of maternal kin related individuals within groups (e.g. matrilines) and this has been documented for several species (e.g. *M. fuscata*, Itani, 1975; *M. mulatta*, Sade *et al.*, 1976; *M. sylvanus*, Ménard and Vallet, 1993a). It is largely due to female philopatry, that macaque groups can be considered as stable social entities. The subgrouping phenomenon described in some studies (*M. nemestrina*, Caldecott, 1986) probably represents groups engaged in a process of fission rather than a species specific group structure characteristic (Oi, 1990). In addition, since the existence of matrilines governs the demographic and genetic structure of groups, the study of their characteristics (number and size) in relation to patterns of group fission are of considerable importance to understanding the microevolutionary processes determining population structure in macaques. Depending on group size and demographic regime within populations, we can expect variations in the number and size of matrilines within groups (Melnick and Kidd, 1983) and thus in the distribution of the degree of relatedness within groups. Moreover, depending on the level of within-group competition, differences in reproductive success between matrilines may emerge, leading to variations in their size.

When a group reaches a maximal size or when competition between matrilines is high, groups are likely to fission into two or more smaller new independent groups. Group fissions represent the only way for females to disperse. Although they are relatively rare and discrete events, they nevertheless play a major role in the dynamics of groups or populations, contributing to a re-distribution of female genes and (potentially) relatedness. Unfortunately however, few fissions have been described, mainly due to the need for long-term field studies of groups in which individuals and kin-relatedness are both known.

Various factors have been specified as being responsible for the occurrence of fission in macaque groups, including sexual competition among males in provisioned (Furuya, 1969) or wild groups (Yamagiwa, 1985; Dittus, 1988), female competition for food (Dittus, 1988; Sugiyama and Oshawa, 1982), or a loss of cohesion between females leading them to spread into different new groups (Oi, 1988; Ménard and Vallet, 1993a; Okamoto and Matsumara, 2001). In all cases, fissions constitute non-random sampling processes within a population, and occur mainly between matrilines. In addition, fissions may occur along lines of dominance relationships (Dittus, 1988).

Frequency of fission is difficult to ascertain, but some information is available. Provisioned groups appear to fission more frequently than wild groups, possibly related to their higher rates of increase in group size and thus their faster attainment of critical maximum size. Furuya (1969) observed five fissions in a total of 12 group-years in provisioned Japanese macaques, indicating that one might expect any one group to fission once every 2.4 years. Ober *et al.* (1984) observed two fissions in provisioned *M. mulatta* groups after a 7-year study of one focal group; i.e. one fission every 3.5 years. In wild *M. fuscata* groups, fissions occurred less frequently (2 over 11 group-years, ie once per 5.5 years, Maruhashi, 1982), whereas in a longer-term study of wild *M. fascicularis* (3 groups over 12-years), only a single fission event was recorded (van Noordwijk and van Schaik, 1999). In *M. sinica,* Dittus (1988) observed 29 wild groups over 5 to 16 years and estimated the occurrence of fission of any one group to be once every 52 years.

Group size at fissioning is highly variable depending on environmental conditions. Fissioning at group sizes of 22-54 has been reported for animals living under wild and non-provisioned conditions (Table 1, *M. sinica*, Dittus, 1988; *M. mulatta*, Melnick and Kidd, 1981; *M. fuscata*, Maruhashi, 1982; Oi, 1988; *M. fascicularis*, van Noordwijk and van Schaik, 1999; *M. maurus,* Okamoto and Matsumara, 2001), whereas corresponding values for provisioned groups can be as high as 138 to 158 individuals (range 48-275) (*M. mulatta*, Chepko-Sade and Sade, 1979; *M. fuscata*, Furuya, 1969; Koyama, 1970). Group size at fissioning can also vary considerably even within the same population. In *M. fascicularis* for example, one group reached nearly

Table 1. Group size and matriline characteristics before and after group fission.

Species	*Environmental conditions*	*Mean group size before fission*	*Mean group size after fission*	*Mean number of matrilines before fission*	*Mean size of matrilines before fission*	*Mean number of matrilines after fission*	*Mean size of matrilines after fission*	*Source*	*Study duration*
M. mulatta	Free-ranging	138 (48-275) N=5	63 (18-164) N=7	20.4 (8-37) N=5	4.1 (1-20) N=11	10.6 (1-30) N=11	4.7 (1-17)	1	20 years
M. fuscata	Provisioned	158 (140-180) N=5	79 (15-150) N=10					2	12 years
M. fuscata	Provisioned	163 N=1		24 N=1	4.3 (1-13)	14.5 (14-15) N=2	4.5 (1-17)	3	13 years
M. fuscata	Habituated non provisioned	35 (22-47) N=2	22 (15-32) N=4					4	5 years
M. fuscata	Wild	54 (N=1)	20 (15-25) N=2					5	2 years
M. sinica	Wild	41 (34-47) N=2	?	7.7 (6-10) N=3	2.7 (1-6) N=3	3.8 (1-6) N=6	3.0 (1-6)	6	16 years
M. mulatta	Wild	29 N=1	14 (10-18) N=2					7	2 years
M. fascicularis *	Wild	About 30 N=1	About 15 N=2					8	12 years

Table 1. Con td.

Species	*Environmental conditions*	*Mean group size before fission*	*Mean group size after fission*	*Mean number of matrilines before fission*	*Mean size of matrilines before fission*	*Mean number of matrilines after fission*	*Mean size of matrilines after fission*	*Source*	*Study duration*
M. maurus	Wild	43 N=1	18.5 (16-21) N=2	9 N=1	2.9 (1-9)	4.5 (4-5) N=2	2.9 (1-9)	9	11 years
M. sylvanus **	Free-ranging	141 (89-227) N=8	63 (13-185) N=12			8.1 (3-19) N=8	3.9 (1-15)	10	20 years
M. sylvanus	Free-ranging	132	64 (37-94) N=2	24 N=1	2.4 (1-5)	13 (10-15) N=2	2.2 (1-5)	11	13 years
M. sylvanus	Wild	88 N=1	29 (13-50)	12 N=1	3.2 (1-9)	4 (1-6) N=3	4.6 (1-9)	12	11 years

The range of sizes is given within brackets.
1: Chepko-Sade and Sade, 1979, 2 : Furuya, 1968, 1969 ; 3: Koyama, 1970; 4: Maruhashi, 1982 ; 5 : Oi, 1988 ; 6 : Dittus 1988 ; 7: Melnick and Kidd, 1981; 8: van Noordwijk and van Schaik, 1999; 9: Okamoto and Matsumara, 2001;. 10 : Küster and Paul, 1997 ; 11 : Prud'homme, 1991 ; 12: Ménard and Vallet, 1993a.
**: data not comparable with our data: The number of matrilines is probably underestimated because a matriline is considered as a whole even if its connector female has dead. The size of matrilines is probably also underestimated because matriline size only included females ≥4 Years. *: data drawn from van Noordwijk and van Schaik (1999, Figure 1, p.109).

50 individuals and another one exceeded 30 individuals without fissioning, while a third group fissioned at about 30 individuals (Figure 1, P. 109, van Noordwijk and van Schaik, 1999). Fission generally leads to a reduction in group size by a factor of 2.

Little information is available about the characteristics of matrilines within any given group, and even data on size and number are limited. What is available has been compiled and presented in Table 1. Studies of wild macaque populations are rarely long enough to enable ancestor females belonging to the oldest generation to be included, since such females are usually already dead when studies begin. Moreover, the social links between daughters are weakened when the matriarch is dead and the probability of splitting of the lineage then becomes higher (Chepko-Sade and Olivier, 1979). For these reasons, in our own studies of Barbary macaques, we have defined a matriline (or lineage) as being composed of females descending from the same, surviving matriarch. Males are excluded. The mean number of matrilines within groups before fission is much greater for provisioned (20-24) than for wild groups (8-9), and this is, at least in part, related to differences in group sizes. Matrilines are larger on average in provisioned (4.1-4.9 females) than in wild groups (2.7-2.9 females) before fission, reflecting more favourable demographic conditions in a captive/semi-captive environment. Moreover, provisioned groups contain some large matrilines composed of more than ten females. After fission, the number of matrilines in the new groups is reduced by a factor of 2 while mean matriline size does not change.

Although probably non-representative for the species, provisioned groups provide useful experimental situations that allow examination of the potential effect of demographic variation on the dynamics and distribution of relatedness within groups. When the rate of increase in group size is rapid due to favourable demographic conditions (provisioned groups), the mean interval between group fissions is shorter and matrilines are composed of many females, potentially leading to a high mean degree of relatedness within groups.

Effects of dynamics on the distribution of genes

The effects of male dispersal can be detected at different levels within macaque populations reflecting either broad distribution of their genes or localised gene transmission according to reproductive tactics. The proportion of nuclear genetic diversity found in a single population generally exceeds 90% of the overall species diversity (*M. nemestrina*, *M. fascicularis*, *M. mulatta*, *M. sinica*; Melnick and Hoelzer, 1992). Moreover, migrations have been found to be responsible for 50% of gene frequency changes in all the free-ranging rhesus groups at Cayo Santiago (Ober *et al.*, 1984). In addition, in accordance with male dispersal occurring mainly locally, neighbouring groups are more genetically similar to each other than are geographically distant groups (*M. fuscata,* Nozawa *et al.*, 1982; *M. mulatta,* Melnick, 1987) and the degree of relatedness between matrilines is highest among adjacent groups (*M. fascicularis,* de Ruiter and Geffen, 1998).

Female philopatry leads to low levels of within-population mtDNA diversity and large between-population differences (Melnick and Hoelzer, 1992). At the group level, the degree of relatedness in *M. fascicularis* is higher among adult females (0.14, the level of cousins), than among adult males (-0.10, de Ruiter and Geffen, 1998). Group fission (occurring primarily along lines of matrilineages) has been shown to be responsible for 13% of the total changes in gene frequency of Rhesus groups at Cayo Santiago, partly due to lineal effects, but also because fission results in increased male migration during and following the event (Ober *et al.*, 1984). Group cohesion apparently depends on the mean degree of relatedness among group members such that groups become candidates for splitting when their average relatedness falls below a particular level, such as second cousin level as observed in rhesus macaques (Chepko-Sade and Olivier, 1979). Fissions occurring along matrilines result in an increase in relatedness within new groups compared with that in the initial group, while the genetic divergence between the new groups increases (*M. mulatta,* Chepko-Sade and Olivier, 1979; Olivier *et al.*, 1981).

Group dynamics and the distribution of relatedness in Barbary macaques

In wild populations of Barbary macaques, groups vary in size from 13 to 88 individuals, all of them presenting a multimale-multifemale structure and a more or less balanced adult sex ratio (Ménard and Vallet, 1996; for a review see also Ménard, 2002). In one case, a group displayed a one-male structure for one year before switching to a classical multimale-multifemale structure as a result of male immigration (Ménard and Vallet, 1993a, 1996). In rocky, mountainous terrain, social groups split frequently into subgroups of small size, temporarily losing their spatial cohesion although maintaining their social links (Ménard *et al.,* 1990).

Male dispersal

Male dispersal is mostly linked to the mating season. In the wild, about 50% of males live in their natal group after reaching sexual maturity and may stay until at least 9 years of age (Ménard and Vallet, 1996). Under free-ranging (semi-captive) conditions, natal migrations are most common around the time of puberty (3-4 years old). However, only one-third of all males leave their natal group and variation in migration rates seems to be related to variation in adult sex ratio (Küster and Paul, 1999). When they leave their social group, emigrant males tend to integrate immediately into neighbouring groups without solitary periods (Küster and Paul, 1999; Ménard and Vallet, 1996). Nevertheless, Mehlman (1986) reported that one male in the wild had walked a distance of 8km on the snow as a solitary individual; he was probably migrating. Such a distance is in accordance with a transfer between two neighbouring groups, considering that home ranges in the Ghomoran group were estimated to be about seven square kilometres (Mehlman, 1989). In most cases males seem to migrate alone. In only one case did we observe two males jointly following one of our focal groups in Algeria (Ménard and Vallet, 1996). We also noted that male dispersal was more or less equally distributed between emigrations and immigrations, with the result that groups were able to maintain their social structure (Ménard and Vallet, 1996).

Group fission

Frequency of fission

Studies on free-ranging groups in Salem provide an interesting long-term example of the dynamics of fission. At Salem 8 group fissions occurred over a 20 year study period (Küster and Paul, 1997). If groups that were removed from the population soon after a group fission are excluded, a total of 7 groups were observed, making 67 group-years of observation in total. Overall, the probability of fission occurring in any given group is therefore about once every 8.4 years. The founder group however, fissioned four times within a 3.5 years period after transfer to the enclosure, probably reflecting a period of adaptation to new conditions. If this initial period is excluded, we found 59 group-years of observation and a probability for a group to fission once every 14 years (calculated from Figure 1, p 947, in Küster and Paul, 1997). Our own observations of two groups in Algeria for 9 and 11 years respectively, yielding a total of 20 group-years of observation, resulted in only one group fission.

Group size at fissioning and matriline characteristics in the resulting new groups

Social conditions change within groups as group size increases. In our studies in Djurdjura (Algeria) we observed a period of instability within the focal group with the formation of temporarily separated subgroups of various compositions (Ménard and Vallet, 1993a). During this period, matrilineages showed occasional intralineal splitting causing, for example, juveniles to be separated from their mother for one or a few days. The process of group fission facilitated male transfer between the new groups and neighbouring groups, leading to an increase in male migration within the population. Group size before fissioning was on average around 140 individuals for free-ranging groups while the only group fission observed in the wild occurred when the group reached 88 individuals (Table 1 and 2). In addition, we observed a pre-fissioning process in another wild group composed of 80 individuals in the Middle Atlas (Morocco). Fission of Barbary macaque groups resulted in the formation of two or three new independent groups. The mean size of these new groups was around 63 individuals in a free-ranging colony (calculated from Figure 1, p 947, in Küster and Paul, 1997) and 29 in the wild (Ménard and Vallet, 1993a). This regulatory process leads wild populations to be composed of groups mostly including 30 to 50 individuals with an average of 44 individuals (Figure 1).

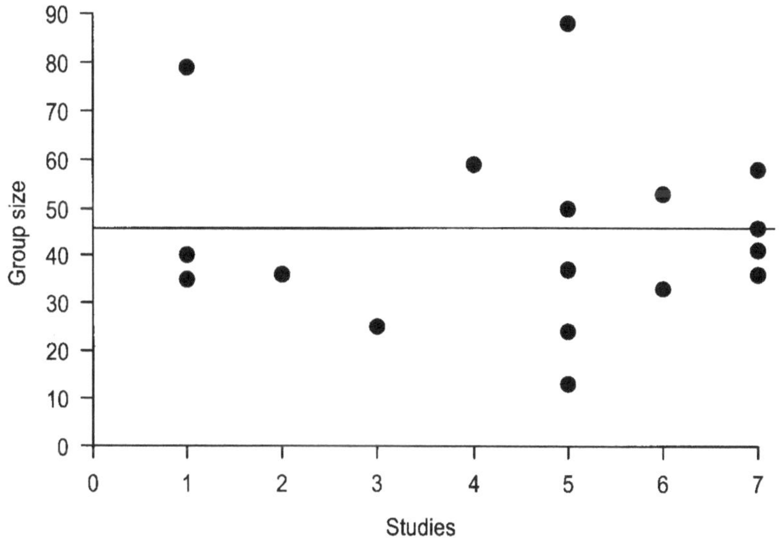

Figure 1. Size of wild Barbary macaque groups according to studies conducted by [1] Ménard, 2002; [2] Taub, 1978; [3] Deag, 1974; [4] Mehlman, 1989; [5, 6] Ménard and Vallet, 1993, 1996; [7] Ménard *et al.*, 1990

Before fission, the Djurdjura group contained 12 matrilines (Ménard and Vallet, 1993a) whereas there were 24 in the free-ranging group of Kintzheim colony (France), studied by Prud'Homme (1991). After fission, the mean number of matrilines within the new groups was also twice as high in free-ranging (8-13) compared to wild groups (4), a difference probably related to group size (Table 1). The mean number of females within matrilines of Algerian groups was relatively stable from year to year (Table 2), the most dramatic changes following the death of a matriarch that led to the creation of numerous new small matrilines (case of Ula group, Table 2). Although the Akfadou group was composed of as many matrilines as the Djurdjura group, fission did not occur (Table 2). Matrilines tended to be larger in the Djurdjura groups than in the Akfadou group with the consequence that matrilines with more than 8 females were more frequent. This possibly reflected the better demographic conditions in Djurdjura than in Akfadou, with a higher adult female and infant survival (Ménard and Vallet, 1993b). The mean number of females per lineage after fission appears greater in the wild (4.6 females) than in the free-ranging colony (3.9 females) studied by Küster and Paul (1997) although in the latter, matrilines only included females ≥4 years (Küster and Paul, 1997). If we also count only females ≥4 years in the Algerian wild groups, matrilines in the wild appear smaller (2.9 females) than in the free-ranging colony. Furthermore, the rather small mean values (2.2) drawn from genealogies in Prud'Homme (Figure 1, p. 12, 1991) may possibly be biased, since the genealogical bonds of the females were not known during the first 5 years of the study.

Table 2. Variation of group size and matriline characteristics.

Djurdjura groups				
Years	*1983-1988*	*1989-1993*		
Group name	Initial group	Smart group	Lorette group	Ula group
Group size	38-88	49-50	24-38	13-34
Number of matrilines	12-14	6-11	5	1-5
Mean annual size of matrilines	2.1-4.1 (1-12)	2.7-6.8 (1-8)	3.2-6 (4-13)	3.2-18 (1-18)

Akfadou group	
Years	*1983-1990*
Group size	33-53
Number of matrilines	10-11
Mean annual size of matrilines	1.6-3.8 (1-8)

Range of matriline size is given within brackets.

In all but one case where genealogies were known, group fissions occurred along matrilines and complete matrilines joined the new groups (Prud'Homme, 1991; Ménard and Vallet, 1993a; Küster and Paul, 1997).

Dispersal patterns and the distribution of relatedness

The modalities of the dispersal of males that join neighbouring groups is reflected in the correlation between geographic distance and genetic distance, based on nuclear microsatellite markers for groups within the population of Djurdjura National Park in Algeria (von Segesser *et al.*, 1999). Moreover, due to female philopatry, one could expect a greater relatedness among adult females than among adult males belonging to the same group. In addition, due to group fission modalities which separate matrilines, one could expect an increase in the degree of relatedness within the new groups after fission. In other respects, group fission may result from a low degree of relatedness within the initial group. In Barbary macaques, such data are only available from the wild groups that were studied for several years in Akfadou forest and Djurdjura National Park in Algeria.

The Akfadou group was studied for 9 years (1983-1991) and contained between 33 and 53 individuals. The Djurdjura groups have been studied for 11 years (1983-1993). The initial group of Djurdjura contained 38 individuals in 1983. After it reached 88 individuals in 1988, it fissioned resulting in the formation of three new groups of 13 (Ula group), 24 (Lorette group) and 50 individuals (Smart group, Ménard and Vallet, 1993a).

Our genetic study was conducted using blood samples from individuals in our focal groups from populations of Akfadou and Djurdjura (Ula group excluded due to insufficient samples). The specific conditions of DNA extraction and amplification are described elsewhere (Lathuillière *et al.*, 2001; Lathuillière, 2002). Analyses were performed using 15 microsatellite markers with a mean expected heterozygozity (*He*) of 70% (Figure 2). Ten of them (D1S548, D2S1326, D3S1768, D4S243, D5S820, D6S493, D7S2204, D8S1106, D10S1432, D16S420) were previously available and tested by us for the first time in *M. sylvanus* (see also, Lathuillière, 2002) while Barbary specific primers were designed for five loci (D1S207, D7S503, D11S925, D17S791, D18S536) according to Lathuillière *et al.* (2001) and Lathuillière (2002).

The degree of relatedness (R) was calculated following the formula of Queller and Goodnight (1989). Mean R values of supposed non-kin pairs (-0.05; N=541) and known mother-infants pairs (0.53; N=21) were close to theoretical values. The mean relatedness among adult males and among adult females within groups was close to non-kin values in both Akfadou and Djurdjura groups before fission (Figure 3). However in the initial Djurdjura group, the relatedness among females was slightly but significantly higher than that found among males, whereas this was not the case in the Akfadou group (Figure 3). The fission of the Djurdjura group resulted in a significant increase in female-female relatedness within the new groups. Mean R reached the level of cousins in Smart group and exceeded the level of half-siblings in Lorette group (Figure 3). These high degrees of relatedness remained identical during the five years following the fission.

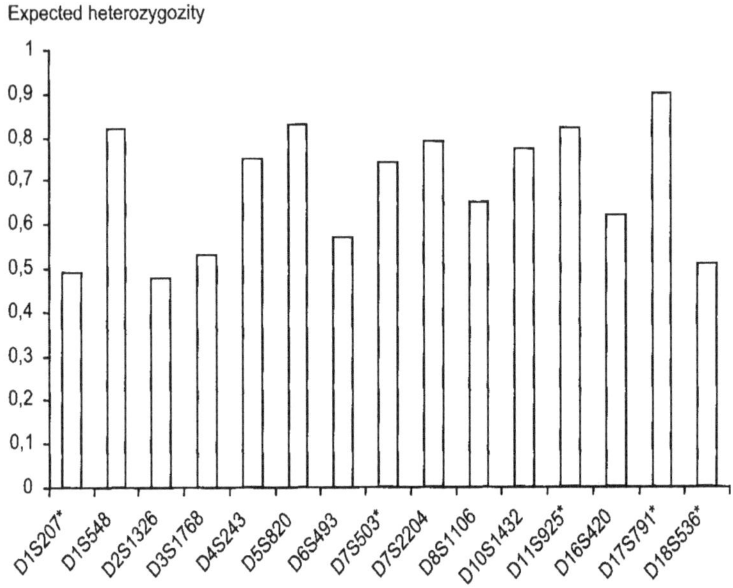

Figure 2. Expected heterozygosity of fifteen microsatellite markers. *: loci amplified with Barbary macaque specific primers (see also, Lathuillière *et al.*, 2001; Lathuillière, 2002).

In contrast, male-male R remained close to the values of non-kin individuals although slightly increasing in the Smart group. This fission effect on the distribution of kinships among groups probably explains the significant genetic divergence found between groups in the Djurdjura population (microsatellites markers, von Segesser *et al.*, 1999; allozyme markers, Scheffrahn *et al.*, 1993).

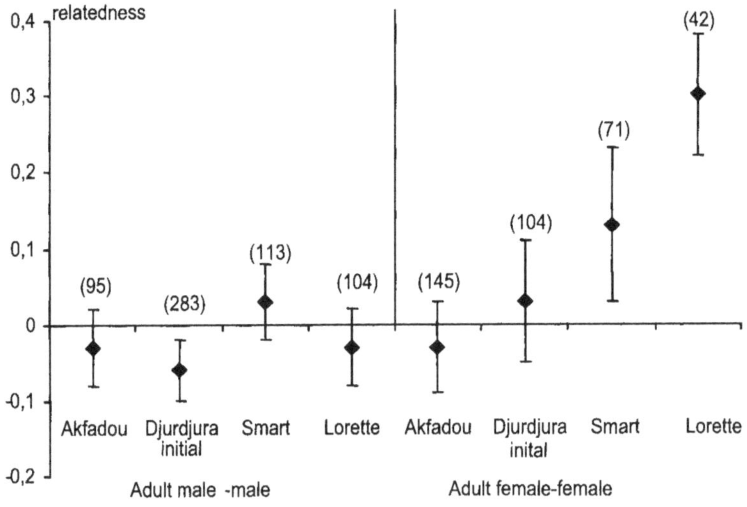

Figure 3. The effect of group fission on the mean relatedness among adult males and adult females. The number of pairwise comparisons is given in brackets.

Discussion

Patterns of male dispersal clearly differ between Barbary macaques and the other macaque species, especially with regard to the age at natal migration. Although male Barbaries remain in their natal group into adulthood and may reproduce there, they probably avoid mating with females from their related matrilines. In free-ranging groups, Paul and Küster (1985) showed that males living in their natal groups avoided mating with closely related females. We hypothesise that incest avoidance is facilitated in large groups composed of many matrilines and conversely, that males tend to migrate earlier when groups contain few matrilines and potentially more related females. This is supported by the fact that the maximum critical size of Barbary macaque groups before fissioning appears to be much larger than that of other macaque species in the wild and that they also contain a greater number of matrilines. One explanation for the ability of Barbary macaques to live in/tolerate larger groups than other species may be related to differences in the distribution of food resources. In contrast to species such as *M. fascicularis* and *M. fuscata* which feed on clumped defensible resources, the diet of Barbary macaques is mainly composed of widely and uniformly distributed items such as herbaceous leaves and seeds (Ménard and Vallet, 1988). Moreover, our own observations of wild populations of Barbaries have shown that increases in group size neither lead to an increase in aggressive behaviour nor to an increase in day-length travels (unpublished). Further clarification of this issue requires additional studies examining the relationship between group size and number of matrilines and the age of males at natal migration in a large sample of groups in different species in natural habitats. Indeed, it is not clear whether these different patterns actually reflect species-specific characteristics or are simply linked to the small number of groups studied in each species.

Group size (both pre- and post-fissioning) is larger in Barbary macaques than in at least three other species (*M. fascicularis, M. fuscata* and *M. maurus*, Table 1). Thus, Barbary macaques seem to be able to tolerate the coexistence of a greater number of matrilines within larger groups. Moreover, matrilines within groups near to the fissioning process are larger in Barbary macaque than in other species. Whether or not this reflects species differences in fission characteristics *per se* or results from differences in demographic regimes of the various species, however, remains unclear. Indeed, the largest matriline size was found in groups of the rapidly expanding population of Djurdjura, which was undergoing an intrinsic annual increase rate of 15% (Ménard and Vallet, 1993b). In this population, matriline sizes approximate those of provisioned groups whereas in another Barbary group in a population with a much lower rate of increase (Afkadou, 4% per year), matriline sizes approximate those of wild groups of other species (Ménard and Vallet, 1993b).

Contrary to our hypothesis, female philopatry does not necessarily imply a high relatedness among females within the same group or a higher degree of relatedness than between males. The low mean R observed among female Barbary macaques could result from genetic divergence between numerous matrilines belonging to the same group. In contrast, de Ruiter and Geffen (1998) reported a high mean relatedness among females in five *M. fascicularis* groups. The divergence between the two studies possibly results from a smaller number of matrilines (1-3) within *M. fascicularis* focal groups, but could also reflect the fact that studies were carried out at different stages in group history (prefissioning period, newly-formed groups for example) rather than any species-specific trend. Indeed, our study on Barbary macaques indicates that the mean female-female R within groups may vary from non-kin to half-sibling levels depending on their history (Lathuillière, 2002, Lathuillière *et al.*, 2004). Our study however does not confirm the threshold of second cousin below which group fission should occur (Chepko-Sade and Sade, 1979). Indeed, the cohesion of the observed Barbary groups both in Akfadou and Djurdjura has persisted for at least 9 years despite a mean R near that of unrelated individuals. Could this threshold vary among species? We suggest that in a highly tolerant species such as the Barbary macaque, individuals are able to display more relaxed relationships towards unrelated individuals than the rhesus macaque which is categorised as a despotic species (Thierry *et al.*, 2000).

Matriline characteristics and fission frequency as well as male mating patterns are key factors determining the level of relatedness within groups, all of which are likely to depend on demographic conditions. Furthermore, ecological conditions (clumped *vs.* widely spread resources for example) and the resuting degree of competition may influence group sizes and the number of unrelated matrilines within groups. In order to clarify the interrelationship between these different parameters, it would be interesting to conduct field studies on species living in different ecological conditions. Unfortunately, this is difficult to investigate in primates. Theoretical approaches may then be of help. Chesser (1991), for example, modelled the influence of philopatry and existence of lineages on population genetic structure of social mammals. More recently, Lefebvre *et al.* (2003) have employed a simulation model to test the effects of the dynamics of the social groups and of matrilines on population structure, hence relatedness. They were able to show that for comparable population growth rates, populations with high survival rates of adult females produce few large matrilines whereas populations with high survival rates of juveniles produce many small matrilines within groups. Subsequent use of analytical models has shown that an increase in natality favours large numbers of matrilines and that their probability of extinction decreases (Caron-Lormier *et al.*, 2006). Thus, the continuing development of analytical models should allow us to address questions that previously would have necessitated difficult field studies involving long-term observations on the influence of demographic conditions on the structure of groups and populations.

In the context of biological conservation, especially in the case of an endangered species such as the Barbary macaque, it is of crucial importance to be able to predict the effect of demographic variation on the distribution of genes within a population, and to estimate the inbreeding risk in small patches or refuges resulting from habitat fragmentation. Analytical models of the different interactions between demography, group dynamics, matriline characteristics (number, size and also depth i.e. the number of generations), and the distribution of genes provide a promising way of improving our understanding of the evolutionary processes occurring within populations and to be able to predict their future.

Acknowledgements

We are grateful to Keith Hodges and John Cortes for inviting us to contribute to this volume and especially Keith Hodges for his help in improving the manuscript. We also thank Dr. W. Scheffrahn and Dr. R.D. Martin for initiating N.M. to genetic techniques fifteen years ago in the Institute of Anthropology of Zürich. Research on Barbary macaques received financial supports from the Ministry of Environment (comity EGPN, France), the CNRS and the University of Rennes1.

References

Caldecott JO (1986) *An ecological and behavioral study of the pig-tailed macaque.* Karger, Basel

Caron-Lormier G, Masson J-P, Ménard N and Pierre JS (2005) A branching process, its application in biology: influence of demographic parameters on the social structure in mammal groups. *Journal of Theoretical Biology* **238** 564-574

Chepko-Sade BD and Olivier TJ (1979) Coefficient of genetic relationship and the probability of intragenealogical fission in *Macaca mulatta. Behavioral Ecology and Sociobiology* **5** 263-278

Chepko-Sade BD and Sade DS (1979) Patterns of group splitting within matrilineal kinship groups: a study of social group structure in *Macaca mulatta. Behavioral Ecology and Sociobiology* **5** 67-86

Chesser RK (1991) Gene diversity and female philopatry. *Genetics* **127** 437-447

Deag JM (1974) *A study of the social behavior and ecology of the wild Barbary macaques, Macaca sylvanus*. PhD thesis, University of Bristol.

de Ruiter JR and Geffen E (1998) Relatedness of matrilines, dispersing males and social groups in long-tailed macaques (*Macaca fascicularis*). *Proceedings of the Royal Society of London* **265** 79-87

Dittus WPJ (1988) Group fission among wild toque macaques as a consequence of female resource competition and environmental stress. *Animal Behaviour* **36** 1626-1645

Furuya Y (1968) On the fission of troops of Japanese monkeys I Five fissions and social changes between 1955 and 1966 in the Gagyusan troop. *Primates* **9** 323-350

Furuya Y (1969) On the fission of troops of Japanese monkeys II. General view of troop fission of Japanese monkeys. *Primates* **10** 47-69

Itani J (1975) Twenty years with Mount Takasaki monkeys. In *Primate utilization and conservation* pp 101-125 Ed G Berment and DG Lindburg. John Wiley and Sons, London

Koyama N (1970) Changes in dominance rank and division of a wild japanese monkey troop in Arashiyama *Primates* **11** 335-390

Küster J and Paul A (1997) Group fission in Barbary macaques (*Macaca sylvanus*) at Affenberg Salem. *International Journal of Primatology* **18** 941-966

Küster J and Paul A (1999) Male migration in Barbary macaques (*Macaca sylvanus*) at Affenberg Salem. *International Journal of Primatology* **20** 85-106

Lathuillière M (2002) *Influence de la fragmentation des populations et des stratégies de reproduction et de dispersion sur la diversité génétique des populations de magots (Macaca sylvanus) [Influence of populations fragmentation and reproductive and dispersal strategies on the genetic diversity of Barbary macaque populations (Macaca sylvanus)]*. PhD thesis, University of Tours.

Lathuillière M, Crouau-Roy B, Petit E, Scheffrahn W and Ménard N (2004) Influence of group fission on gene dispersion in the Barbary macaque (*Macaca sylvanus*) *Folia Primatologica* **75** 171-172

Lathuillière M, Ménard N and Crouau Roy B (2001) Sequence conservation of nine Barbary macaque (*Macaca sylvanus*) microsatellite loci: implication of specific primers for genotyping. *Folia Primatologica* **72** 85-88

Lefebvre D, Ménard N and Pierre JS (2003) Modelling the influence of demographic parameters on group structure in social species with dispersal asymmetry and group fission. *Behavioral Ecology and Sociobiology* **53** 402-410

Maruhashi T (1982) An ecological study of troop fissions of Japanese monkeys (*Macaca fuscata yakui*) on Yakushima Island, Japan. *Primates* **23** 317-337

Mehlman PT (1986) Male intertroup mobility in a wild population of the Barbary macaque (*Macaca sylvanus*). Ghomaran Rif Mountains. Morocco. *American Journal of Primatology* **10-** 67-81

Mehlman PT (1989) Comparative density, demography, and ranging behavior of Barbary macaques (*Macaca sylvanus*) in marginal and prime conifer habitats. *International Journal of Primatology* **10** 269-292

Melnick DJ (1987) The genetic consequence of primate social organization: a review of macaques, baboons and vervet monkeys. *Genetica* **73** 117-135

Melnick DJ and Hoelzer GA (1992) Differences in male and female macaque dispersal lead to contrasting distributions of nuclear and mitochondrial DNA variations. *International Journal of Primatology* **13** 379-393

Melnick DJ and Kidd KK (1981) Social group fission and paternal relatedness. *American Journal of Primatology* **1** 333-334

Melnick DJ and Kidd KK (1983) The genetic consequences of social group fission in a wild population of rhesus monkeys (*Macaca mulatta*) *Behavioral Ecology and Sociobiology* **12** 229-236

Melnick DJ, Pearl M and Richard AE (1984) Male migration and inbreeding avoidance in wild rhesus monkeys. *American Journal of Primatology* **7** 229-243

Ménard N (2002) Ecological plasticity of Barbary macaques (*Macaca sylvanus*). *Evolutionary Anthropology* **11** 95-100

Ménard N (2004) Do ecological factors explain variations in macaque social organisation? *How societies are built: the macaque model?* pp 237-266 B Thierry, M Singh and W Kaumanns. Cambridge University Press, Cambridge

Ménard N and Vallet D (1988) Disponibilités et utilisation des ressources par le magot; (*Macaca sylvanus*) dans différents milieux en Algérie [Available resources and use by Barbary macaques (*Macaca sylvanus*) in different habitat types in Algeria]. *Revue d'Ecologie.(Terre et Vie)* **43** 201-250

Ménard N and Vallet D (1993a) Dynamics of fission in a wild Barbary macaques group (*Macaca sylvanus*). *International Journal of Primatology* **14** 479-500

Ménard N and Vallet D (1993b) Population dynamics of *Macaca sylvanus* in Algeria: an 8-year study. *American Journal of Primatology* **30** 101-118

Ménard N and Vallet D (1996) Demography and ecology of Barbary macaques (*Macaca sylvanus*) living in two different habitats. In *Evolution and ecology of macaque societies* pp 106-131 Eds JE Fa and DG Lindburg. Bambridge University Press, Cambridge

Ménard N, Hecham R, Vallet D, Chikhi H and Gautier-Hion A (1990) Grouping patterns of a mountain population of *Macaca sylvanus* in Algeria-A fission-fusion system? *Folia Primatologica* **55** 166-175

Mori A and Watanabe K (2003) Life history of male Japanese macaques living on Koshima Islet *Primates* **44** 119-126

Nozawa, K, Shotake T, Kawamoto Y and Tanabe Y (1982) Population genetics of Japanese monkeys.II. Blood protein polymorphism and population structure. *Primates* **23** 252-271

Ober C, Olivier TJ, Sade DS, Schneider SM, Cheverud J and Buettner-Janush J (1984) Demographic components of gene frequency change in free-ranging Macaques on Cayo Santiago. *American Journal of Physical Anthropology* **64** 223-231

Oi T (1988) Sociological study on the troop fission of wild japanese monkeys (*Macaca fuscata yakui*) on Yakushima island. *Primates* **29** 1-19

Oi T (1990) Population organization in wild pig-tailed macaques (*Macaca nemestrina)* in West Sumatra. *Primates* **31** 15-31

Okamoto K and Matsumura S (2001) Group fission in Moor macaques (*Macaca maurus*). *International Journal of Primatology* **22** 481-493

Okamoto K, Matsumura S and Watanabe K (2000) Life history and demography of wild moor macaques (*Macaca maurus*): Summary of ten years of observations. *American Journal of Primatology* **52** 1-11

Olivier TJ, Ober C, Buettner-Janusch J and Sade DS (1981) Genetic differentiation among matrilines in social groups of rhesus monkeys. *Behavioral Ecology and Sociobiology* **8** 279-285

Paul A and Küster J (1985) Intergroup transfer and incest avoidance in semifree-ranging Barbary macaques (*Macaca sylvanus*) at Salem (FRG). *American Journal of Primatology* **8** 317-322

Prud'Homme J (1991) Group fission in a semifree-ranging population of Barbary macaques (*Macaca sylvanus*) *Primates* **32** 9-22

Queller DC and Goodnight KF (1989) Estimating relatedness using genetic markers. *Evolution* **43** 258-275

Sade DS, Cushing K, Cushing P, Dunaif J, Figueroa, A., Kaplan JR, Lauer C, Rhodes D and Schneider J (1976) Population dynamics in relation to social structure on Cayo Santiago. *Yearbook of Physical Anthropology* **20** 253-262

Scheffrahn W, Ménard N, Vallet D and Gaci B (1993) Ecology, demography and population genetics of Barbary macaques in Algeria. *Primates* **34** 387-400

Sugiyama Y (1976) Life history of male Japanese macaques. *Advances in the study of behaviour* **7** 255-284

Sugiyama Y and Ohsawa H (1982) Population dynamics of Japanese monkeys with special reference to the effect of artificial feeding. *Folia Primatologica* **39** 238-263

Taub DM (1978) *Aspects of the biology of the wild Barbary macaque (Primates: Cercopithecinae, Macaca sylvanus L. 1758): biogeography, the mating system and male-infant interactions.* PhD thesis, University of California,

Davis.

Thierry B, Iwaniuk AN and Pellis SM (2000) The influence of phylogeny on the social behaviour of macaques (Primates: Cercopithecidae, genus *Macaca*) *Ethology* **106** 713-728

van Noordwijk MA and van Schaik CP (1985) Male migration and rank acquisition in wild long-tailed macaques (*Macaca fascicularis*). *Animal Behaviour* **33** 849-861

van Noordwijk MA and van Schaik CP (1999) The effects of dominance rank and group size on female lifetime reproductive success in wild long-tailed macaques, *Macaca fascicularis*. *Primates* **40** 105-130

von Segesser F, Ménard N, Gaci B and Martin RD (1999) Genetic differentiation within and between isolated Algerian subpopulations of Barbary macaques (*Macaca sylvanus*): evidence from microsatellites. *Molecular Ecology* **8** 433-442

Yamagiwa J (1985) Socio-sexual factors of troop fission in wild japanese monkeys (*Macaca fuscata yakui*) on Yakushima island, Japan. *Primates* **26** 105-120

Yoshimi I and Takasaki H (2003) Long distance mobility of male Japanese macaques evidenced by mitochondrial DNA *Primates* **44** 71-74

Zhao QK (1994) Mating competition and intergroup transfer of males in Tibetan Macaques (*Macaca thibethana*) at Mt. Emei, China. *Primates* **35** 57-68

Four-year old sub-adult barbary macaques playing in the decidous oak forest of Akfadou, Algeria. Copyright D. Vallet

Disease transmission in Barbary and other macaques: risks and implications for management and conservation

PE Honess[1], M Pizarro[2], NN Sene[3] and SE Wolfensohn[1]
[1] *Dept of Veterinary Services, University of Oxford, Parks Road, Oxford, OX1 3PT, UK*
[2] *Gibraltar Veterinary Clinic, 36 Rosia Road, Victualling Yard, Gibraltar*
[3] *Microbiology Dept, Laboratories of Clinical Pathology and Public Health, Gibraltar Health Authority, Gibraltar*

Introduction

The health of animals and habitats, particularly those of specific conservation interest, has become of increasing concern to conservationists (e.g. Deem *et al.*, 2001; Laffety and Gerber, 2002). With mounting pressure on global resources, some animals have become restricted to progressively smaller areas of natural habitat, creating not only ecological but also social pressures and increasing the opportunity for spread of disease both within and across species barriers, to and from Man (Karesh *et al.*, 2002). As a result, a new interdisciplinary approach has been called for to tackle these emerging challenges to health, that of Conservation Medicine (see Aguirre *et al.*, 2002).

Primates are capable of acting as reservoirs for a number of infectious pathogens that may have very serious consequences for humans and, conversely, disease transmission from humans can significantly threaten endangered populations of primates (Jones-Engel *et al.*, 2001; McCallum and Dobson, 1995). The study of these diseases in primates in wild and semi free-ranging populations as well as in captivity will increase our understanding of their origin, evolution and epidemiology, thus facilitating the development of treatments and the assessment of their consequences for threatened primate populations (Hunt and Desrosiers, 1994; Wolfe *et al.*, 1998).

Macaques have been an important model for the study of human diseases for many years and a significant body of literature on disease research in this genus now exists. Historically this dates back to the 1960's and the use of large numbers of wild caught rhesus macaques (*Macaca mulatta*) in the search for, and development of, a vaccine for polio. In more recent times, rhesus and several other macaque species (e.g. stumptailed: *M. arctoides*, long-tailed macaques: *M. fascicularsis*; pig-tailed: *M. nemestrina*) have been used with varying degrees of success as models for research into such diseases as tuberculosis, HIV, malaria and conditions such as asthma. The maintenance of large numbers of primates in captivity for biomedical and behavioural research has led to considerable detailing of diseases that affect them in the captive environment (Good *et al.*, 1969; Fowler, 1993; Bennett *et al.*, 1998). However, relatively little is known of diseases affecting macaques under natural conditions in the wild.

This chapter will outline the major diseases found in macaques particularly those which can be transmitted to humans. Specific reference is made to the Barbary macaques (*Macaca sylvanus*) on Gibraltar and some recent data on the health status of this unique semi-free ranging colony are presented. The monitoring and consequences of disease in primates for the management of small and threatened populations are discussed, particularly in the context of conservation goals for threatened species.

Main diseases affecting macaques

There is a range of diseases that affect macaques just as there is for other species. Table 1 lists the most important macaque diseases together with their significance, manifestation in macaques and standard recommended control methods. Details on the range and prevalence of enteric parasites that occur in macaques are given elsewhere (e.g. Fowler, 1993; Jones-Engel *et al.*, 2004).

Table 1. Potential zoonotic diseases in macaques.

Infectious agent	*Relative zoonotic significance*	*Nature of the infection in the macaque*	*Control method and transmission*
Herpesvirus simiae (B virus)	+++	Benign in the macaque but health and safety significance as >70% fatal in humans. Many wild caught macaques are seropositive with latent infection and will be lifelong carriers, virus shedding increases with stress	Screen, cull or protection to reduce human exposure
Simian immuno-deficiency virus (SIV)	++	Will cause clinical disease and death in the macaque	Screen and cull. Spread to human via blood and body fluids, seroconversion in human but clinical significance in human not fully known

Table 1. Contd.

Infectious agent	*Relative zoonotic significance*	*Nature of the infection in the macaque*	*Control method and transmission*
Simian T-lympho-trophic virus (STLV-1)	++	Often subclinical in the macaque, may get immunosupression	Seroconversion in human but clinical significance in human not fully known
Simian retrovirus (SRV-1, -2, -3)	++	Often subclinical in the macaque, may get immunosupression	Seroconversion in human but clinical significance in human not fully known
Simian foamy Virus (SFV)	+	Very common: 89.5% seroprevalence in *Macaca fascicularis* in Bali (Jones-Engel *et al.*, 2005)	Seroconversion in human but clinical significance in human not fully known
Hepatitis A	++	Juveniles susceptible, virus shed in faeces, then may be seropositive with no infection	If virus is shed, faeces may be source of infection for humans. Monitor by serology and virus isolation. Control by reducing exposure to faecal material, vaccination
Filoviruses (Ebola, Marburg)	+++	Various strains with differing pathogenicity to macaques and humans. No latent infection, once seropositive, virus is eliminated (Schou and Hansen, 2000)	Quarantine for 31 days, check for clinical signs, reduce contact during this time
Measles	++	Outbreaks seen occasionally, usually of human origin	Vaccination possible, control exposure of susceptible macaque to infected human
Tuberculosis	+++	Chronic clinical disease and eventually death. Spread in respiratory droplets and in faeces	Screening at post mortem or TB testing (but high incidence of false positives and negatives). High risk of infection spreading from human to macaque as well as *vice versa*
Melioidosis	+	Endemic in some areas, particularly seen after the wet season. May remain latent for prolonged time	Of significance in immunocompromised humans

Table 1. Contd.

Infectious agent	*Relative zoonotic significance*	*Nature of the infection in the macaque*	*Control method and transmission*
Bacterial enteric pathogens (e.g. shigella)	+++	Depending on the bacteria and the body condition and stress level of the animal, and whether already immuno-suppressed by concomitant viral infection, may get severe diarrhoea and death or may become chronic carrier and shed infection intermittently	Routine faecal screening to check on incidence. Can be spread to humans and *vice versa*. Control by reducing contact with faecal material, use of disinfectants and hygienic practices (Kennedy *et al.*, 1992). Control insect vectors which will spread infection (Cohen *et al.*, 1991; Weil *et al.*, 1971).
Enteric parasites (e.g. *Balantidium coli*)	++	Frequently present at low levels but can cause disease and death if high level of infection especially if immuno-suppressed by presence of viral infection	Routine faecal screening to check on incidence. Can be spread to humans and vice versa. Control by reducing contact with faecal material, use of disinfectants and hygienic practices (Owen, 1992).

Disease in Barbary macaques

The Barbary macaques on Gibraltar have been the subject of detailed record keeping for many years. Data on disease in the population are however patchy with no records being available for significant periods of time. It is however known that during the period from 1936 to 1981, nine different causes of death (for approximately 34% of reported deaths; in 66% no cause was reported) were recorded among the macaques and detailed by Fa (1984) as: fungal dermatophytic disease (7 cases); hypothermia (2); pneumonia (2); laryngo-tracheo bronchitis (1); coronary thrombosis (1); gastro-enteritis (15); inguinal hernia (1); accidental electrocution (3); injuries (4) and 'destroyed' (15). In more recent times there has been extensive screening of the Gibraltar population and over the last few years it has been serologically tested for various zoonotic diseases. Table 2 and Table 3 give details of the testing and results from recent screening. All macaques which were blood tested also received an intra-palpebral (intradermal, using the upper eyelid site) injection of tuberculin (0.1ml [1500iu] intradermal tuberculin mammalian, human isolates) to test for tuberculosis; none of which resulted in the identification of a positive result.

Table 2. Results of analyses of blood samples from different groups of Gibraltar Barbary macaques (*Macaca sylvanus*) carried out at the Gibraltar Health Authority Hospital laboratory in Gibraltar, 2000/2001.

Macaque group (approx size)	*No. sampled*	*Hiv Abs 1and2/p24*	*HBsAg*	*Anti-HCV*	*HAVT*	*HAV IgM*
Middle Hill (60)	26	All negative	Negative	Negative	24 Positive	Negative
Farringtons (60)	30	Negative	Negative	Negative	24 Positive	Negative
Anglian Way (45)	15	Negative	Negative	Negative	13 Positive	Negative
Prince Phillip (64)	13	Negative	Negative	Negative	10 Positive	Negative
Apes Den (32)	12	Negative	Negative	Negative	6 Positive	Negative

(Key: HIV Abs 1and2/p24 - HIV antibodies. HBsAg - Hepatitis B virus surface antigen. Anti-HCV - Hepatitis C virus antibodies. HAVT - hepatitis A viral antibody test. HAV IgM - Hepatitis A virus, immunoglobulin M.).

The only disease of significance for which there were positive results was hepatitis A. However none of the macaques that tested positive had serological evidence of previous exposure to the disease (IgM [immunoglobulin M] antibodies not detected) and no animal showed clinical signs of illness attributable to hepatitis A. The main clinical significance of this test result is that any macaque with an active clinical infection may be shedding the virus and this may have zoonotic consequences.

Most cases of viral hepatitis seen in Man are caused by one of the following: hepatitis A, hepatitis B or hepatitis C (nonA, nonB) (Zuckerman and Howard, 1979; Hollinger and Dienstag, 1985); all result in systemic disease primarily involving the liver with the possibility of fulminant liver failure. Infection with these viruses has been detected in a number of nonhuman primates including macaques (Zuckerman *et al.*, 1978; Anan'ev *et al.*, 1984; Lankas and Jensen, 1987; Balayan, 1992; Bukh *et al.*, 2001; MacDonald *et al.*, 2000). Evidence of hepatitis infection in people associating with nonhuman primates is documented in the literature, indicating that there may be a risk of zoonotic disease (Kovacs *et al.*, 1974; Deinstag *et al.*, 1976; Schiller and Ochs, 1993). The question as to whether hepatitis A, B or C virus infection among the 200 – 250 Barbary macaques (*Macaca sylvanus),* living semi-wild in the Nature Reserve on the Rock of Gibraltar, is a public health risk has been examined by measuring seroprevalence among a sample of animals (Sene, unpublished) (see Table 2).

Using the Enzyme Linked Fluorescent Assay (ELFA) technique for the detection of Hepatitis A virus IgG (HAVIgG) and IgM (HAVIgM) antibodies, Hepatitis B virus surface antigen (HBsAg), and Hepatitis B surface antibodies (anti-HBs); and the Enzyme immunoassay (EIA) for the detection of Hepatitis C virus antibodies (anti-HCV), it was found that troops were equally seropositive with HAVIgG, but the other hepatitis virus seromarkers were undetectable in the serum samples. There was no significant difference in seroprevalence between males and females, but there was a higher incidence in macaques that were 5 years of age or older. HAVIgM was not detected and as HAV is not a chronic infection, this suggests that the virus is not endemic. As such, young susceptible macaques do not have any antibodies against HAV and can therefore still become infected and could subsequently transmit infection to humans.

The source of this infection is unknown. It is possible that it may be of human origin, either coming directly from contact with tourists or via the food supply, although Balayan (1992) has also suggested that spontaneous HAV infection has occurred in captive non-human primates. This possibility could be addressed by determining the amino acid sequence of the HAVIgG isolated from positive macaques. Future correlation of HAV infection in the macaques with that of HAV infection in humans that come into contact with the animals would assist in determining any potential public health risks. Faecal screening for presence of virus would indicate whether there is active infection with shedding of infectious agents which would be a public health risk. In the meantime, active immunisation of susceptible young may be indicated in order to reduce the level of infection and transmission. However this is a complex strategic decision involving assessment of the costs and benefits (to people and animals) as well as public relations outcomes and therefore cannot be considered as mandatory in all instances.

Table 3. Results of analyses of blood samples from different groups of Gibraltar Barbary macaques (*M. sylvanus*) carried out at the Biomedical Primate Research Centre (BPRC), Rijswijk, The Netherlands, 2003.

Macaque group (approx size)	*No. sampled*	*Alpha-Herpes virus*	*SIV*	*SRV*	*STLV*
Middle Hill (60)	54	negative	negative	negative	negative
Farringdon's (60)	33	negative	negative	negative	negative
Anglian Way (45)	3	negative	negative	negative	negative
Prince Philip (64)	5	negative	negative	negative	negative
Apes Den (32)	4	negative	negative	negative	negative

Enteric bacteria in Barbary macaques

Faecal testing has also been conducted over the last 6 years on over 200 samples, of which 21 cultured salmonella and 34 cultured campylobacter: in all cases the animals were clinically healthy. These organisms appear to present the major risk of zoonosis transmission from the macaques although unfortunately there has been no study to ascertain if there is an increased incidence of gastroenteritis among visitors to the macaque sites and there are no details on serotypes.

Ectoparasites in Barbary macaques

Pedicinus albidus (sucking lice) have been identified in around 50 % of all macaques trapped on Gibraltar during the period 2004-5 (Cohn *et al.*, in prep.). These parasites are species-specific and not thought to be able to survive on humans. Infestation of macaques is short term but may cause discomfort from pruritus and secondary infection. Animals are treated every six weeks with oral ivermectin presented in various types of foods, but it has proved difficult to eradicate the lice. This is mainly due to the random nature of treatment as it is impossible to ensure that all animals receive the medication. Additionally, macaques that are trapped for blood sampling (for testing) or identification (microchipping and tattooing) purposes, are also routinely injected with subcutaneous ivermectin (0.2mg/kg).

None of the diseases discussed above has been judged to have resulted in any of the deaths among the Gibraltar macaques, since those reported by Fa in 1984. The main "natural" cause of death between 2000 and 2005 has been road traffic accidents with on average 5-10 macaques killed in this way per annum. In a number of isolated cases animals have died or been euthanised as a result of: Diabetes mellitus (1 case); osteosarcoma of the tibia (1); hind limb paralysis in one month old macaque with severe head injuries (1); euthanasia due to severe injury (3); pneumonia (1 juvenile: histology results indicated the possible cause of death to be morbillivirus, probably measles virus); and chronic debilitating arthritis (1).

Monitoring diseases and their transmission between macaques and humans

Disease prevention is clearly preferable to subsequent treatment. Additionally, since a number of infectious diseases of primates are potentially zoonotic, a programme of health screening not only improves primate health and welfare and assists in the prevention of disease, but also contributes to the occupational health programme of the host institute or zoo as well as to public health programmes where macaques and people interact in wild, or semi free ranging situations (see Wolfensohn and Honess, 2005).

Macaques can carry a number of particularly pathogenic diseases, such as salmonellosis and tuberculosis, and the UK Advisory Committee on Dangerous Pathogens has made particular recommendations with respect to simian herpesvirus (*Herpesvirus simiae*, B virus) and simian retroviruses. A health screening programme should be drawn up for every facility holding primates which should take account of the source of the animals, the use to which they are put (e.g. breeding, experimental, zoo or nature reserve with public access), and the resources available for testing.

Both animals and people working in close contact with them should, where possible, have their health regularly monitored. Where testing is carried out it should be appropriate to the epidemiological nature of the disease in terms of the timing and frequency of test and type of test used. Issues relating to the specificity and sensitivity of tests need to be accounted for in the interpretation of results as in a number of instances it is possible to record either false positive or false negative results (Wolfensohn and Gopal, 2001; Wolfensohn and Honess, 2005).

Risks and implications

Disease and small population management

As well as extrinsic factors such as management decisions to cull based on real or perceived threats to human health, the presence of an infectious disease can affect the viability of a population in two intrinsic ways: through increased mortality and reduced reproduction (loss of animals of reproductive age, loss of breeding condition) (McCallum and Dobson, 1995). Sophisticated models can be produced that predict the viability of small populations faced with catastrophic events. Fa and Lind (1996) produced just such a model for the Gibraltar macaques using available data on the population's demography and productivity. In the face of a major threat in the form of a disease epidemic, the model illustrated that this population would be at high risk. As a result, the authors concluded that disease prevention should receive the highest priority in the Gibraltar population's management, specifically pointing to the value of detailed record keeping, the collection of comprehensive baseline data on obesity, parasite load and pathogens, rabies vaccination and the performance of post-mortem examination (Fa and Lind, 1996). In more general terms, when monitoring a population, records should be kept on reproductive performance, vaccination history, health screening results (indicating pathogen load), post-mortem examinations, diet and nutrition, trauma, non-infectious disease (e.g. the incidence of neoplasia) and social and psychological factors such as group membership or transfer, significant behavioural changes and dominance status. While accurate modelling and record keeping are important, it is also necessary to be able to assess the impact of disease in natural populations. Standard veterinary approaches may not provide the full picture of the impact, and therefore some researchers propose carefully designed manipulative studies (that are acceptable for threatened species) on the most important parasites and pathogens that may affect a population (McCallum and Dobson, 1995).

Some diseases may increase mortality in specific age-sex classes and therefore have different consequences for the viability of the population. They may specifically affect more vulnerable members of the population - the oldest (e.g. neoplasia) and the weakest (e.g. shigellosis). Whilst these individuals may not have played a direct role in reproduction and their loss may not substantially affect the viability of the population, nevertheless they may play an important social role and their loss may affect group cohesion and stability. Other diseases such as those that are fight-related (e.g. rabies, secondary infections) or sexually transmitted (e.g. *Staphylococcus*) may increase mortality predominantly among younger individuals affecting both their own, as well as their potential offspring's, future recruitment to the population. Non-mortality related reduction in reproduction can be caused by temporary or permanent sterility (e.g. bacterial vaginitis caused by *E. coli* or *Staphylococcus*: Doyle *et al.*, 1991) or loss of breeding condition resulting from disease (e.g. endo- and ectoparasitic loads). Precautions can be taken against disease and its spread by:

a) Quarantining any introductions before release, allowing a full health screen and, if necessary, a course of vaccinations (Cunningham, 1996).

b) Periodic health screen of the whole population, or a randomly selected sample which is appropriate both to the assumed infection rate and the probability of detection of the infection (see Wolfensohn and Lloyd, 2003). Logistical and other practical constraints restrict the choice of strategy so that, for example, it may be highly desirable, practical and financially viable to regularly screen a whole laboratory colony (which is conditioned to regular capture and sedation), but there will be few situations in small free-ranging populations where any strategy other than the examination of random or *ad hoc* individuals is possible.

c) Monitoring of endoparasites and potentially also antibodies or antigens of some systemic infections through faecal and urine analysis (Wolfe *et al.*, 1998).

d) Where possible the isolation, followed by diagnosis and treatment of diseased individuals.

e) Culling may be necessary as a last resort where testing conclusively indicates untreatable disease that poses a significant threat to human health or the survival of the rest of the population.

As indicated above, the Gibraltar population of *Macaca sylvanus* is relatively disease free (see section 2.1) possibly due to its isolation from conspecifics and the fact that they appear to have come originally from a small group of macaques imported from Morocco and Algeria (Modolo *et al.*, 2005) As a result of probability, taking a small number of animals from a population reduces the likelihood that the removed animals may have diseases common to the population as a whole.

The reasons for the low mortality among the Gibraltar macaques include:

- the virtual disease-free status of the population;
- year round provisioning resulting in reduced food competition and no shortages;
- availability of emergency veterinary care for individuals that would otherwise die or be culled; and
- regular parasite treatment.

These low measures of disease and mortality mean that the viability of the Gibraltar population is much improved and the main problem has become one of controlling macaque numbers. On the other hand, the population remains highly vulnerable to potential exposure to a novel disease to which it has no immunity. The introduction of such a disease, most likely from tourists or the local resident human population, would have catastrophic consequences. The macaques are a great tourist attraction, with most of the approximately one million people visiting the Nature Reserve every year specifically wanting to see the monkeys. Some tour operators encourage their clients to interact with the monkeys and the high degree of physical contact with tourists results in numerous bite incidents. With the macaque population expanding, there are regular incidences of animals entering the town area and raiding dustbins and houses. This could be a serious public health problem since faecal contamination by the macaques can potentially result in the transmission of zoonotic diseases. When these animals become urban residents they are deemed too much of a public health risk and nuisance and they have to be culled, as it has proved impossible to relocate these animals to any other nature park.

In Gibraltar, one of the main management problems is that of disease risk and nuisance to the human population when the macaques become urbanised. Management and conservation strategies should therefore centre on the control of macaque numbers and demographics to help prevent group fragmentation and the resultant movement of animals into built-up areas, particularly as the carrying capacity of the Gibraltar Nature Reserve is limited by its relatively small size (about 97 hectares). This should be done using a combination of contraception and selective culling to maintain family structure and appropriate behaviour while still allowing some marginalization and transfer of males to encourage gene flow between groups.

It should not be overlooked that such direct management may be very stressful for the animals as are a range of other factors including increased competition due to ecological pressure (Aguirre *et al.*, 2002), aggressive competition for provisioned resources (Fa and Lind, 1996), disturbance by human visitors (Chamove *et al.*, 1988) and direct intervention such as translocation (Honess *et al.*, 2004). Exposure of the animals to such sources of stress should be minimised as it is known that raised levels of stress may result in immunosuppression that in turn increases disease susceptibility (Chellman *et al.*, 1992; Maule and VanderKooi, 1999; Honess and Marin, 2006; Honess *et al.*, 2005).

Human-macaque interactions

Disease transmission between conspecifics in the wild can take place in a number of ways, including fight-related injuries (e.g. bites), venereally, vertically from the mother to the foetus or infant, through vectors (e.g. biting insects) and from livestock. However, even in wild populations and depending on the infectious agent, physical contact between humans and macaques presents one of the most likely routes of transmission. In some situations, a degree of contact is unavoidable due to the high level of management required (e.g. laboratory and some zoo contexts) and in others (e.g. some tourist attractions such as exist at Asian temple sites and Gibraltar) contact, although strongly discouraged, may still take place. Measures can be taken to reduce the risk of infection during interactions through the wearing of appropriate clothing (including barrier clothing) and staff and visitor education on risks, standard operating procedures and "Dos and Don'ts". Details of interactions between humans and wild or semi-free ranging macaques are given elsewhere (e.g. Fa, 1984, 1989, 1992; Fuentes, this volume).

Disease transmission may not only result from direct contact, such as a bite or scratch, with an infected animal but also from contact with contaminated body fluids, blood or waste matter (e.g. Zwartouw *et al.*, 1984;

Weigler, 1992). Despite high levels of disease prevalence and interaction between humans and primates in the wild and captivity, transmission of some of the most potentially dangerous diseases from primates to humans remains rare (B virus: Engel *et al.*, 2002; Huff and Barry, 2003).

Conservation issues

As indicated in the introduction, wild primate populations are coming under ever-increasing pressure from accelerated development and habitat erosion that are a natural product of the human population increases currently being witnessed across the majority of primate range countries (Harcourt and Parks, 2003). The degree of concern for the survival of many primate populations is reflected in the substantial body of literature on the subject much of which highlights the plight of the primates (e.g. Cowlishaw, 1999; Chapman and Peres, 2001; Harcourt, 2002) and some, while pointing to ethical considerations, emphasises the resource consequences for the local human populations of human-primate conflict situations (e.g. Hill, 2000; Wieczkowski, 2005). Macaques have not been immune from these pressures and populations of both threatened (e.g. Sprague, 2002; Priston, 2005) and more widely distributed species (e.g. Malik and Johnson, 1994; Pirta *et al.*, 1997) are affected. Rarely is permanent separation practical and therefore one or more of a range of mechanisms designed to reduce damage, loss and human-primate contact are usually applied to enable continued safe coexistence (e.g. Strum, 1994).

Conservation issues relating to human threats to primate health

In addition to safeguarding human livelihoods and resources, an important aspect of securing safe co-existence is to limit the potential for disease spread between the local human and wild nonhuman primate populations. As described earlier in this chapter disease transmission can be in either direction: human-animal or animal-human, and either may constitute a threat to the continued survival of the nonhuman primate population. Wallis and Lee (1999) describe a range of diseases, including tuberculosis, enteroviruses and respiratory infections, that were most likely transmitted from local human populations or their domestic animals to wild primate populations (particularly chimpanzees), sometimes with devastating consequences. These consequences may be further magnified in small, isolated populations with low genetic diversity. The genetic distinctiveness and low genetic diversity in all Barbary macaque populations not only makes the case for the conservation of each (Modolo *et al.*, 2005) but also highlights their vulnerability to catastrophic disease events. Fa and Lind (1996) have modelled the potential impact on the Gibraltar population of Barbary macaques of an event such as the introduction of a disease to which the animals have no prior exposure or natural resistance; illustrating that in such a small population, the effects could be "catastrophic". Measures need to be taken to minimise the risk of these sorts of catastrophic events, particularly in relation to disease transmission.

Efforts to study and earn tourist revenue from primate populations present challenges in controlling disease transmission by increasing contact incidents (e.g. habituation and following of animals for research) that are additional to those posed through conflict or incidental contact with local human communities (Wallis and Lee, 1999). It is clear that, in addition to the recommendations concerning health screening outlined above, there are a number of simple measures that can be taken that will contribute to reducing disease risk and other negative consequences of human presence; these include restriction of human-animal contact/proximity, health screening, and high sanitation and waste disposal standards. However, arguably the most important element of any strategy to reduce risk is a programme of appropriate education for local communities, managing authorities, local healthcare providers, veterinary and research staff, and not least the tourists (e.g. Wallis and Lee, 1999; Else and Pokras, 2002; Grossberg *et al.*, 2003).

Conservation issues relating to primate threats to human health

Of the 16 recognised species of macaque, 3 are classified as Endangered, 2 as Threatened and 6 as Vulnerable (Rowe, 1996). *Ex Situ* captive breeding programmes are likely to play an increasingly important role in conserving these species and concerns about the zoonotic and pathogenic nature of a number of the most important diseases (particularly B virus) may threaten such programmes. Conservationists may be left with stark choices between screening and indiscriminate culls, both scenarios resulting in the unacceptable loss of vital genetic resources. Where the legislative framework permits it may be possible to take decisions not to cull populations/colonies of macaques that test positive, or are not unequivocally negative. An example is the decision not to cull the 26 B virus infected Japanese macaques (*Macaca fuscata*) at Launceston, Australia in 2000 (Shanley, 2000, 2002). B virus infection of humans through contact with macaques is very rare, considering the prevalence of this infection in macaque populations throughout many Asian countries and in research facilities (National Research Council, 2003). This may have important implications for present and future efforts to conserve threatened macaques and affects the maintenance of *ex situ* populations of macaques that may be of conservation importance. Regulating and animal health authorities can make powerful decisions about the fate of captive, semi free-ranging or commensal populations that may present a perceived risk to human health such as those that **may** be B virus positive. The apparent difficulty in transmission, particularly outside laboratory settings, should therefore form part of the informed context in which future decisions should be made about whether to cull infected groups.

There are no reports of infections of humans by macaques outside of laboratories or other scientific primate facilities. This is probably partly due to the low degree of contact between staff and animals in such facilities which greatly reduces the risk of transmission of the disease. However it should be noted that it is in laboratory environments that macaques are most likely to be stressed or immuno-suppressed and it is in this state that they are most likely to shed the virus (Chellman *et al.*, 1992; Wolfensohn and Gopal, 2001). Combining this with knowledge of the B virus status of the animals and the high degree of staff/animal contact in these facilities increases the likelihood of correct diagnosis of the disease when transmitted to a human. In non-laboratory environments, particularly in the case of semi-free ranging or wild populations, a reduced index of suspicion may reduce the accuracy of diagnosis thereby depressing the number of cases recorded where infection occurs. Where there is contact between the public and such macaque populations (e.g. Gibraltar - Barbary macaque; Indonesia - long-tailed and pig-tailed macaques) that may be of uncertain disease status this contact should be minimised. Where possible an appropriate programme of screening should be implemented and efforts made to select for animals which test negative for the most important pathogens. Any plans to supplement populations by introducing new individuals or any group translocations should include the health screening of these animals (Wolfe *et al.*, 1998; Tutin *et al.*, 2001) coupled with an awareness that the stress associated with transport may result in immunosuppression and increased susceptibility to opportunistic infections (Honess *et al.*, 2004), including reactivation of a latent or subclinical infection. When working closely with macaques of unknown health status, suitable precautions should be taken (e.g. wearing appropriate protective clothing) as any unfortunate incidents may jeopardise the continued existence of primate colonies that have a high conservation, education or tourism value.

Conclusion

While it has been shown that macaque populations that are either relatively free from (e.g. Gibraltar macaques), or carrying disease (e.g. Bali long-tailed macaques) can coexist with humans without extensive disease transmission, the implications of even a single outbreak could be catastrophic for the population. The potential cost, in terms of human health, economics and conservation effort of zoonotic disease outbreaks can be enormous and efforts to eradicate pathogens or the animal populations that are their reservoirs can also be costly. Therefore it has been proposed that a better approach is to reduce the risk of disease outbreak by reducing contact between species (in this case humans and macaques) (Karesh *et al.*, 2005). A reduction in the risk of transmission coupled with appropriate surveillance and screening will improve the safety of both threatened primate populations and the human populations with which they temporarily or permanently share their environment.

Acknowledgements

The authors would like to thank Keith Hodges and John Cortes for their invaluable editorial comments on this chapter. We also thank the Government of Gibraltar, The Gibraltar Ornithological and Natural History Society, The German Primate Centre and Institut Scientifique Rabat for funding, facilitating and organising the Calpe 2003 conference which resulted in this collaboration. Specifically thanks to Eric Shaw, Damian Holmes, Dale Laguea and Dr John Cortes (GONHS). We also thank our institutions for their support in the production of this chapter.

References

Aguirre, A.A., Ostfeld, R.S., Tabor, G.M., House, C. and Pearl, M.C. (2002). Conservation Medicine: Ecological Health in Practice. Oxford University Press, New York 407pp.

Anan'ev, V.A, Viazov, S.O., Garanina, N.M., Dorshenko, N.V. and Zhilina, N.N.(1984). Viral hepatitis A and B in anthropoid apes of Moscow Zoo. *Voprosy Virusologii* **29**(4): 434-437.

Balayan, M.S. (1992). Natural hosts of hepatitis A virus. *Vaccine* **10** Supplement 1: S27-31.

Bennett, B.T., Abee, C.R. and Hendrickson, R. (eds.) (1998). Nonhuman Primates in Biomedical Research: Diseases. Academic Press, San Diego.

Bukh, J., Forns, X., Emerson, S.U. and Purcell, R.H. (2001). Studies of hepatitis C virus in chimpanzees and their importance for vaccine development. *Intervirology* **44**(2-3): 132-142.

Chamove, A.S., Hosey, G.R. and Schaetzel, P. (1988). Visitors excite primates in zoos. *Zoo Biology* **7**, 359–369.

Chapman, C.A. and Peres, C.A. (2001). Primate conservation and the new millennium: The role of scientists. *Evolutionary Anthropology* **10**: 16-33.

Chellman, G.J., Lukas, V.S., Eugui, E.M., Altera, K.P., Almquist, S.J. and Hilliard,

J.K. (1992). Activation of B virus (*Herpersvirus simiae*) in chronically immunosuppressed cynomolgus monkeys. *Laboratory Animal Science* **42**: 146-51.

Cohen, D., Green, M. and Block, C. (1991). Reduction of transmission of shigellosis by control of houseflies (*Musca domestica*). *The Lancet* **337**: 993.

Cohn, D.L., Smith, V., Pizarro, M., Jones-Engel, L., Engel, G., Fuentes, A., Shaw, E. and Cortes, J. Pediculosis in *Macaca sylvanus* of Gibraltar. In prep.

Cowlishaw, G. (1999). Predicting the pattern of decline of African primate diversity: An extinction debt from historical deforestation. *Conservation Biology* **13**(5): 1183-1193.

Cunningham, A.A. (1996). Disease risks of wildlife translocations. *Conservation Biology* 10(2): 349-353.

Deem, S.L., Karesh, W.B. and Weisman, W. (2001). Putting theory into practice: Wildlife health in conservation. *Conservation Biology* **15**(5): 1224-1233.

Deinstag, J.L., Davenport, F.M., McCollom, R.W., Hennessy, A.V., Klatskin, G. and Purcell, R.H. (1976). Nonhuman primate-associated viral hepatitis type A. Serological evidence of hepatitis A virus infection. *Journal of the American Medical Association* 236(5): 462-464.

Doyle, L., Young, C.L., Yang, S.S. and Hillier, S.L. (1991). Normal vaginal aerobic and anaerobic bacterial flora of the rhesus macaque (*Macaca mulatta*). *Journal of Medical Primatology* 20: 409-413.

Else, J.G. and Pokras, M.A. (2002). Introduction. In, Aguirre, A.A., Ostfeld, R.S., Tabor, G.M., House, C. and Pearl, M.C. (eds). Conservation Medicine: Ecological Health in Practice. Oxford University Press, New York p. 3-7.

Engel, G.A., Jones-Engel, L., Schillaci, M.A., Suaryana, K.G., Putra, A., Fuentes, A. and Henkel, R. (2002). Human exposure to herpesvirus B-seropositive macaques, Bali, Indonesia. *Emerging Infectious Diseases* **8** (8): 789-795.

Fa, J.E. (1984). Structure and dynamics of the Barbary macaque population in Gibraltar. In: Fa, J.E. (ed.). *The Barbary Macaque: A Case Study in Conservation.* Plenum Publishing, New York. pp. 263-306.

Fa, J.E. (1989). Influence of people on the behaviour of display primates. In: Segal, E.F. (ed.). *Housing, Care and Psychological Wellbeing of Captive and Laboratory Primates*. Noyes Publications, Park Ridge, New Jersey, pp. 270-290.

Fa, J.E. (1992). Visitor-directed aggression among the Gibraltar macaques. *Zoo Biology* **11**: 43-52.

Fa, J.E. and Lind, R. (1996). Population management and viability of the Gibraltar Barbary macaques. In: Fa, J.E. and Lindburg, D.G. (eds.), *Evolution and Ecology of Macaque Societies*. Cambridge University Press, Cambridge. pp 235-262.

Fowler, M.E. (ed.) (1993). *Diseases of Zoo and Wild Animals.* 3rd Edition. W.B.

Saunders, USA.

Good, R.C., May, B.D. and Kawatomari, T. (1969). Enteric pathogens in monkeys. *Journal of Bacteriology* **97**:1048-1055.

Grossberg, R., Teves, A. and Naughton-Treves, L. (2003). The incidental ecotourist: Measuring visitor impacts on endangered howler monkeys at a Belizean archaeological site. *Environmental Conservation* **30**(1): 40-51.

Harcourt, A.H. (2002). Rarity, specialization and extinction in primates. *Journal of Biogeography* **29**: 445-456.

Harcourt, A.H. and Parks, S.A. (2003). Threatened primates experience high human densities: Adding an index of threat to the IUCN Red List criteria. *Biological Conservation* **109**(1): 137-149.

Hill, C.M. (2000). Conflict of interest between people and baboons: Crop raiding in Uganda. *International Journal of Primatology* **21**(2): 299-315.

Hollinger, F.B. and Dienstag, J.L. (1985). Hepatitis viruses. In: Lennette, E.H., Balows, A., Housler, W. and Shadomy, H.J. (Eds.) *Manual of Clinical Microbiology*. American Society for Microbiology, Washington D.C., p. 813-835.

Honess, P. and Marin, C. (2005). Behavioural and physiological aspects of stress and aggression in nonhuman primates. *Neuroscience and Biobehavioral Reviews*. doi:10.1016/j.neubiorev.2005.04.003.

Honess P., Marin, C., Brown, A. and Wolfensohn, S. (2005). Assessment of stress in non-human primates: application of the neutrophil activation test. *Animal Welfare* **14**. In Press.

Honess, P., Johnson, P. and Wolfensohn, S. (2004). A study of behavioural responses of non-human primates to air transport and re-housing. *Laboratory Animals* **38**: 119-132.

Huff, J.L. and Barry, P.A. (2003). B-virus (*Cercopithecine herpesvirus* 1) infection in humans and macaques: Potential for zoonotic disease. *Emerging Infectious Diseases* **9** (2): 246-250.

Hunt, R.D. and Desrosiers, R.C. (1994). Study of spontaneous infectious diseases of primates: Contributions of the Regional Primate Research Centers Program to conservation and new scientific opportunities. *American Journal of Primatology* **34**(1): 3-10.

Jones-Engel, L., Engel, G.A., Schillaci, M.A., Aida Rompis, A., Putra, A., Suaryana, K.G., Fuentes, A., Beer, B., Hicks, S., White, R., Wilson, B. and Allan, J.S. (2005). Primate-to-human retroviral transmission in Asia. *Emerging Infectious Diseases* **11**(7):1028-1035.

Jones-Engel, L., Engel, G.A., Schillaci, M.A., Babo, R. and Froehlich J. (2001). Detection of antibodies to selected human pathogens among wild and pet macaques (*Macaca tonkeana*) in Sulawesi, Indonesia. *American Journal of Primatology* **54**(3): 171-178.

Jones-Engel, L., Engel, G.A., Schillaci, M.A., Froehlich, J., Paputungan, U. and

Kyes, R.C. (2004). Prevalence of enteric parasites in pet macaques in Sulawesi, Indonesia. *American Journal of Primatology* **62**(2): 71-82.

Karesh, W.B., Cook, R.A., Bennet, E.L. and Newcomb, J. (2005). Wildlife trade and global disease emergence. *Emerging Infectious Diseases* **11**(7): 1000-1002.

Karesh, W. B., Osofsky, S. A., Rocke, T. E., and Barrows, P. L. (2002). Joining forces to improve our world. *Conservation Biology* **16**(5): 1432-1434.

Kennedy, F.M., Astbury, J. and Needham, J. R. (1992). Shigellosis due to occupational contact with non-human primates. *Epidemiology and Infection* **110**: 247-254.

Kovacs, J., Marzinsky, P. and Altman, R. (1974). Probable chimpanzee-associated hepatitis A - New Jersey. *CDC Morbidity and Mortality Weekly Report* **23** (43): 372.

Lafferty, K.D. and Gerber, L.R. (2002). Good medicine for conservation biology: The intersection of epidemiology and conservation theory. *Conservation Biology* **16**(3): 593-604.

Lankas, G.R. and Jensen, R.D. (1987). Evidence of hepatitis A infection in immature rhesus monkeys. *Veterinary Pathology* **24**(4):340-344.

McCallum, H. and Dobson, A. (1995). Detecting disease and parasite threats to endangered species and ecosystems. *Trends in Ecology and Evolution* **10**(5):190-194.

MacDonald, D.M, Collins, E.C., Lewis, J.C.M. and Simmonds, P. (2000). Detection of hepatitis B virus infection in wild-born chimpanzees (*Pan troglodytes verus*): Phylogenetic relationships with human and other primate genotypes. *Journal of Virology* **74** (9): 4253-4257.

Malik, I. and Johnson, R.L. (1994). Commensal rhesus in India: The need and cost of translocation. *Revue D'Ecologie (Terre and Vie)* **49**(3): 233-244.

Maule, A. and VanderKooi, S. (1999). Stress induced immune-endocrine interaction. In: Balm, P. (Ed.), *Stress Physiology in Animals.* Sheffield Academic Press, Sheffield, UK, pp. 205–245.

Modolo, L., Salzburger, W. and Martin, R.D. (2005). Phylogeography of Barbary macaques (*Macaca sylvanus*) and the origin of the Gibraltar colony. *Proceedings of the National Academy of Sciences* **102** (20): 7392-7397.

National Research Council (2003). Occupational Health and Safety in the Care and Use of Nonhuman Primates. The National Academies Press, Washington DC

Owen, D.G. (1992). *Parasites of Laboratory Animals.* Laboratory Animal Handbooks No. 12 Royal Society of Medicine Press.

Pirta, R.S., Gagil, M. and Kharshikar, A.V. (1997). The management of the rhesus monkey *Macaca mulatta* and Hanuman langur *Presbytis entellus* in Himachal Pradesh, India. *Biological Conservation* **79**: 97-106.

Priston, N.E.C. (2005). *Crop-Raiding by Macaca ochreata brunnescens in*

Sulawesi: Reality, Perceptions and Outcomes for Conservation. PhD Thesis, University of Cambridge.

Rowe, N. (1996). *A Pictorial Guide to the Living Primates.* Pogonias Press, New York.

Schiller, W.G. and Ochs, A. (1993). Hepatitis A infection caused by occupational contact with primates. *Arbeitsmedizin Sozialmedizin Umweltmedizin* **28** (12): 530-532.

Schou, S. and Hansen, A.K. (2000). Marburg and ebola virus infections in laboratory non human primates: A literature review. *Comparative Medicine* **50**: 108-123.

Shanley, L. (2002). A wonderful day for the Lunceston macaques. *IPPL News* **29**(1): 28. See also http://www.ippl.org/05-02-28-2.html.

Shanley, L. (2000). Zoo monkeys saved from death. *IPPL News* **27**(3): 5. See also http://www.ippl.org/200saved.html.bak.

Sene, N.N. (unpubl.). Seroprevalence of hepatitis A, B and C virus infection among Barbary macaques (*Macaca sylvanus*) in Gibraltar: A public health risk? *The Barbary Macaque: Comparative and Evolutionary Perspectives.* Gibraltar, 5-8 November 2003. Gibraltar Ornithological and Natural History Society, Gibraltar.

Sprague, D. (2002). Monkeys in the Backyard: Encroaching wildlife and rural communities in Japan. In, Fuentes, A. and Wolfe, L.D. (Eds.) *Primates Face to Face: The conservation implications of Human-Nonhuman Primate Interconnections.* Cambridge University Press, Cambridge, UK. p. 254 - 272.

Strum, S.C. (1994). Prospects for management of primate pests. *Revue D'Ecologie (Terre and Vie)* **49**(3): 295-306.

Tutin, C.E.G., Ancrenaz, M., Paredes, J., Vacher-Vallas, M., Vidal, C., Goossens, B., Bruford, M.W. and Jamart, A. (2001). Conservation biology framework for the release of wild-born orphaned chimpanzees into the Conkouati Reserve, Congo. *Conservation Biology* **15**(5): 1247-1257.

Wallis, J. and Lee, D.R. (1999). Primate conservation: The prevention of disease transmission. *International Journal of Primatology* **20**(6): 803-826.

Weigler, B.J. (1992). Biology of B virus in macaque and human hosts: A review. *Clinical Infectious Disease* **14:** 555-67.

Weil, J.D., Ward, M.K. and Spertzel, R.O. (1971). Incidence of *Shigella* in conditioned rhesus monkeys (*Macaca mulatta*). *Laboratory Animal Science* **21**: 434-437.

Wieczkowski, J. (2005). Comprehensive conservation profile of Tana mangabeys. *International Journal of Primatology* **26**(3): 651-660.

Wolfe, N.D., Escalante, A.A., Karesh, W.B., Kilbourn, A., Spielman, A. and Lal, A.A. 1998. Wild Primate Populations in Emerging Infectious Disease Research: The Missing Link? *Emerging Infectious Diseases*

4(2): 149-158.

Wolfensohn, S.E. and Gopal, R. (2001). Interpretation of serological test results for Simian herpes B virus. *Laboratory Animals* **35**: 315-320.

Wolfensohn, S.E. and Honess, P.E. (2005). *Handbook of Primate Husbandry and Welfare*. Blackwell Publishing, Oxford. viii, 168pp.

Wolfensohn, S.E. and Lloyd, M. (2003). Handbook of Laboratory Animal Management and Welfare. 3rd Edition. Blackwell Publishing, Oxford. xii, 416pp.

Zuckerman, A.J. and Howard, C.R. (1979). *Hepatitis Viruses of Man*. Academic Press, London.

Zuckerman, A.J., Thornton, A., Howard, C.R., Tsiquaye, K.N., Jones, D.M. and Brambell, M.R. (1978). Hepatitis B outbreak among chimpanzees at the London Zoo. *Lancet* **2** (8091): 652-654.

Zwartouw, H.T., Macarthur, J.A., Boulter, E.A., Seamer, J.H., Marston, J.H. and Chamove, A.S. (1984). Transmission of B virus infection between monkeys especially in relation to breeding colonies. *Laboratory Animals* **18**: 125-30.

Patterns and context of human-macaque interactions in Gibraltar

A Fuentes
Department of Anthropology, University of Notre Dame, Notre Dame, IN, 46556-5611 USA

Introduction

Nonhuman primates are our closest evolutionary relatives and integral elements in many cultures, mythologies, diets, and scientific paradigms. Specific focus on the multifarious interaction of human and nonhuman primates is relevant in primatological and anthropological studies, and it is possible to view human and nonhuman primates as co-participants in a rapidly escalating realm of ecological and cultural change (Fuentes and Wolfe 2002). Of particular relevance is the situation of interconnection that characterises many macaque monkey (genus *Macaca*) populations; that of temple and/or urban macaque populations and monkey-tourist sites. At many of these locations local and foreign humans come into regular contact with macaques. These macaques have become accustomed (some may argue adapted) to interacting with humans, especially in the context of food provisioning. In these cases mutualism can arise from the food provided to macaques and the potential economic benefits that tourists may provide the local populace (Fuentes and Gamerl, 2005; Fuentes *et al.*, 2005). In addition to mutualism, competition and contests over space and resources can also emerge from human-macaque overlap and interaction.

Close contact and range overlap between humans and macaques introduces a very real and potentially dangerous situation of disease transmission (Engel *et al.*, 2002; Jones-Engel, Engel, Schillaci, Rompis, Putra, Suaryana, Fuentes, Beers, Hicks, White, and Allen, in press). In areas of South and Southeast Asia, temples, with associated forests, have been a refuge for monkeys (genus *Macaca*) due to local religious beliefs and land use patterns (Aggimarangsee, 1992; Fuentes *et al.* 2005; Wheatley, 1999; Zhao, 2005). In some areas foreign and domestic tourists enter the temples and the associated forest to interact with the monkeys and view the temple grounds (Fuentes *et al.*, 2000; Fuentes *et al.*, 2005; Schlotternhausen 2000; Wheatley 1999; Wheatley and Harya

Putra 1994; Zhao, 2005). In these cases, there can be substantial economic benefit to local communities from entrance fees as well as the income added to the local businesses. If aggression and contact occur when tourists come into contact with macaques there are health risks to both species (Engel *et al.* 2002; Fa 1992; Wallis and Lee 1999). Potential health risks to humans include possible viral transmission of simian T cell lymphotropic viruses (STLV), simian retrovirus (SRV), simian foamy virus (SFV) and Herpes B virus in addition to other non-viral pathogens (Engel *et al.* 2002; Jones-Engel *et al.* in press; Wolfe *et al.*, 2004). Macaques, in turn, are potentially at risk to an array of human pathogens including measles, influenzas and other respiratory diseases (Jones-Engel *et al.*, 2001).

As humans and macaques become increasingly sympatric across a broad array of habitats, managing their interactions becomes a need for human populations (Fuentes and Gamerl, 2005; Fuentes *et al.*, 2005). Because of the complexity in patterns of human-macaque overlap, understanding the diverse components of interaction patterns can facilitate the creation of a quantitative dataset from which to effectively assess macaque-human interconnections in their behavioural, epidemiological, ecological, and cultural contexts. This dataset can then be used in the construction of management models and recommendations that explicitly focus on both macaque and human behaviour.

Research on human-macaque overlap and co-ecology in Bali, Indonesia, suggests that human-macaque interactions are frequent and may be influenced by multiple behavioural, ecological, and epidemiological variables (Engel *et al.* 2002; Fuentes *et al.* 2005; Fuentes and Gamerl 2005; Wheatley 1999). However, these data are derived from only one species of macaque (*Macaca fascicularis*) in the context of temple monkey-human associations in Southeast Asia. Many species of macaques co-exist with humans in a diverse array of habitats across Asia, in Northern Africa, and in Gibraltar. To date, there has been very little quantitative study of human-macaque interactions from across these locales. Thus, the unique situation in Gibraltar presents a particularly interesting case study to examine, in depth, human-macaque interaction patterns (Fa 1992; Fa and Lind 1996; O'Leary and Fa 1993). In Gibraltar there is a relatively large population of macaques (>200 individuals), with a well known population history, which has a geographically restricted range neighbouring an urban area of high human density and is visited throughout the year by large numbers of tourists. As the only non-animal park, free ranging macaque population in Europe, Gibraltar presents a valuable opportunity to enhance the database for macaque-human interaction by adding a non-Asian interaction site and data from the only non-Asian macaque, the Barbary macaque (*Macaca sylvanus*), to the emerging understanding of macaque-human interconnections. In addition to extending the overall human-monkey interaction database, information from Gibraltar can also produce a comparative data set for analysing the Asian contexts and contribute to the ongoing management endeavours in Gibraltar itself.

The Barbary Macaques of Gibraltar

The history of the Barbary macaques on Gibraltar has been well reviewed by Fa (1984), O'Leary and Fa (1993), and Shaw and Cortes (this volume), so I will not attempt to provide a thorough historical overview here. However, a brief summary of the provisioning and interaction history is relevant. From the turn of the century until the mid 1990s there were two main groups of macaques in the population at Gibraltar; one at Apes' Den (formerly Queen's Gate) and another at Princess Caroline's Battery. By 1946 both groups were provisioned separately, and by 1970 provisioning for the Princess Caroline's Battery group was moved to Middle Hill (a restricted military reservation area) to reduce the macaques' incursions into Gibraltar town (Fa 1984). From 1972 onward the Middle Hill group had minimal interactions with non-military humans whereas the Apes' Den group has been regularly visited at the site since at least 1936 (Fa 1984). Increased and regular human interactions began in 1960 with the promotion of the Gibraltar "Apes" as a tourist attraction. As tourism increased substantially during the 1980s-90s economic pressures came to bear on the taxi drivers and tour guides ferrying tourists to visit the macaques. This may have led to illegal provisioning by some Gibraltarans at new areas on the Rock of Gibraltar (Perez and Bensusan 2005, Shaw and Cortes, this volume). By the late 1990s this had contributed to a fissioning of extant groups which by 2004 had resulted in the permanent presence of at least 6 groups on the Rock (Shaw and Cortes, this volume). Today the main interaction sites are Apes' Den, Anglian Way/St. Michael's cave, Prince Philip's Arch and the Cable Car station at the top of the Rock. (Map 1) It is clear that interactions with humans have been a substantial aspect in the daily lives of the Gibraltar macaques for several generations and that these interactions impact the social behaviour and ecology of the macaques (Fa 1984; Fa and Lind 1996; O'Leary and Fa 1993; Perez and Bensusan 2005, Shaw and Cortes, this volume).

Throughout this history of interactions some records on the impact of disease have also been kept. In the period of 1936-1944 gastroenteritis appears to have had a substantial mortality impact on the Gibraltar macaque population and pneumonia caused high mortality amongst infants in the late 1980s (Fa 1984; O'Leary and Fa 1993). Both of these cases may reflect

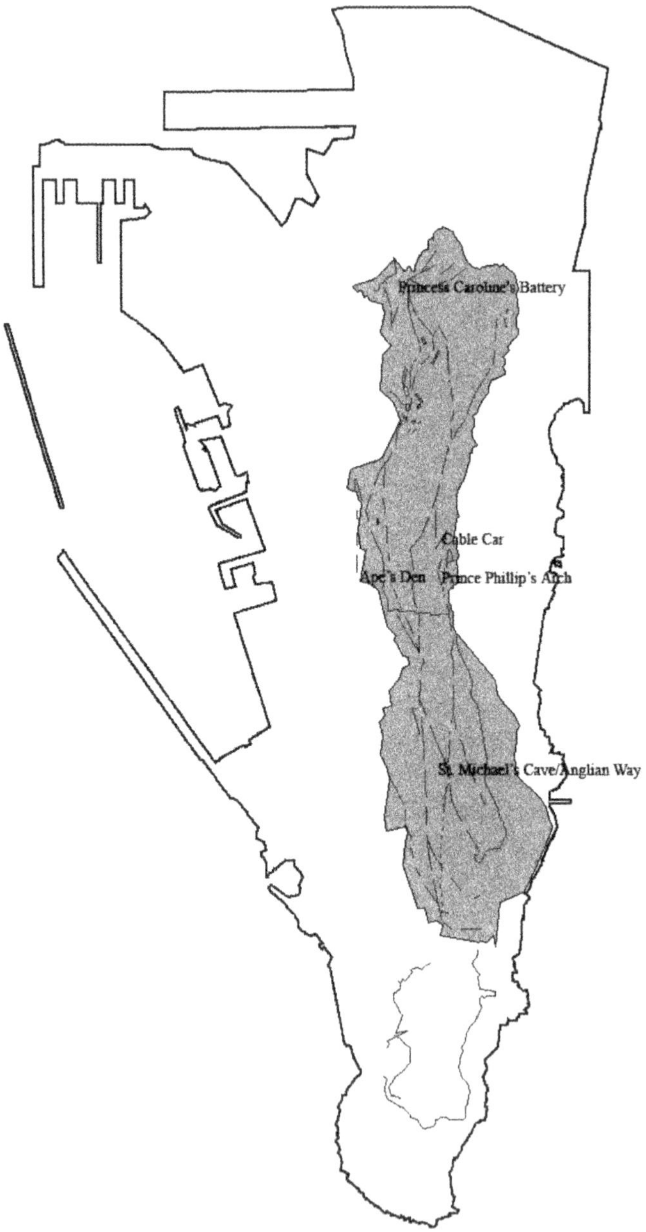

Figure 1. The location of the main interaction sites in the Upper Rock Nature Reserve (reserve area noted in green).

pathogen transmission from humans to macaques. Unpublished observations of intermittent outbreaks of nasal discharge and coughing amongst the macaque groups also suggests that the macaques may be acquiring mild respiratory pathogens from humans (A. Fuentes personal observation and personal communication staff members of Gibraltar Ornithological and Natural History Society (GONHS)). To date there are no published records of a viral pathogen transmission event from the Gibraltar macaques to humans, however the transmission of *campylobacter a*nd a macaque louse to humans may be possibilities (M. Pizarro personal communication).

Barbary macaque-human interactions in Gibraltar

Until recently the only published data on interactions between Barbary macaques and humans came from O'Leary and Fa (1993) (see also Fa 1992). These authors examined Barbary macaque-human interactions in Gibraltar by macaque age/sex classes at the Apes' Den site (then known as Queen's Gate). They found that adult macaques were the age/sex most likely to interact with humans and that adult male macaques were most likely to engage humans by grabbing food or exhibiting "begging" contact. However, they report that infants and juvenile macaques were the age/sex class most commonly observed initiating contact with humans. They also report that subadult male macaques initiated a disproportionately high level of aggressive attacks/threats.

While contributing to our overall understanding of the Gibraltar macaques, this study focussed only on one group during a period when the population size and structure were quite different than today, thus its current applicability is limited.

More recently, preliminary studies by this author's own research team were undertaken to establish a basal assessment of the interactions between macaques and humans in the Gibraltar Nature Reserve during a high tourist season. Data collection design for this project was formulated specifically in the context of the general hypotheses for macaque-human interactions suggested by work with *Macaca fascicularis* in Bali (Wheatley, 1999; Wheatley and Harya Putra, 1994; Gamerl and Fuentes, 2005). Data were collected between June 6th-June 30th, 2004 at four sites in the Gibraltar Nature Reserve: Apes' Den, St. Michael's cave/Anglian way, Prince Philip's Arch, and Cable Car Station. A fifth site, Princess Caroline's Battery/Farringdon's Barracks, also received some data collection, but at a rate lower than the four primary sites, and is not reported here. Each site is a bounded, established location in the reserve with easily defined boundaries and is adjacent to a provisioning location where staff members of GONHS provide food and water on a daily basis for the macaque groups. During observations macaque aggression was categorised into four levels: facial threat (AG1), displace (AG2), chase without contact (AG3), or chase with contact or grab, climb, hit, or bite (AG4). Humans were divided into adult males, adult females, male children (~15 years of age or less), and female children and macaques were classified as adult male, adult female, sub-adult male (75-100% adult size with incomplete canine eruption), sub-adult female (nulliparous and 75-100% adult size), or immature (less than 75% adult size). Fifteen Chi-square tests were used to assess basic patterns in the dataset for this analysis. Due to the number of tests run a Bonferronni correction was used to establish the minimum confidence level of .003 (.05/15 comparisons), so a p value of less than .003 is considered as the indicator of significance here. A Spearman's rank correlation was used to assess the relationship between human and macaque densities and macaque bites directed at humans. For a detailed overview of the methodology see Anderson *et al.* (in prep) and Fuentes (in press).

Summary of overall data

We collected at total of 17,775 minutes (296.25 hours) of data between June 6th and June 30th, 2004. This includes 1020 samples of all occurrence demographic data (10,200 minutes) with 5489 interactions and 505 focal macaque follow samples (7575 minutes) of which 288 (57%) involved one or more interactions between humans and macaques. There were a total of 808 interactions recorded in full detail during focal follows.

Table 1. Summary of the all occurrence interaction data across four sites in Gibraltar.

	Apes' Den site	*Anglian Way/ St. Michael's Cave site*	*Cable Car site*	*Prince Philip's Arch site*	*Total: all sites combined*
Total no. 10 minute all-occurrence samples	271	196	222	271	1020
Avg. no. macaques per sample	13.75	4.74	2.16	15.14	7.344
Avg. no. humans per sample	44.31	67.73	27.65	32.95	37.55
Ratio: macaques to humans at site	1 to 3.22	1 to 14.2	1 to 12.8	1 to 2.18	1 to 5.1
Total interactions	1570	662	290	2957	5489
Contact interactions	974	468	204	2216	3867
Non contact interactions	596	194	86	741	1622
Total no. bites	5	2	13	19	39

Table 1a. Statistical Analyses- All Occurrence sample interactions.

Site	*Interaction rate*	*Contact vs. non-contact intercations*	*Bites*
Apes' Den	.58 per minute, (X^2=.296,p>.05) - as expected		
Anglian Way/ St. Michael's	.34 per minute, (X^2=7.4,p<.01>.001) - as expected		
Cable Car	.13 per minute, X^2=31.3,p<.001 - less than expected		
Prince Philip's Arch	1.1 per minute X^2=58, p<.001 - more than expected		
All sites Combined	Rates differ across sites (X^2=96, p<.001)	More contact than non-contact (X^2=459,p<.001).	Rates differ across sites (X^2=18.3,p<.001)

The four main sites are significantly different in their rates of interaction. Apes' Den does not differ from the expected rate of interaction based on the population mean, nor did Anglian way/St. Michael's. However, Prince Philip's Arch is above the expected interaction rate and Cable Car station had less than the expected rate of interactions. Across all sites combined there are significantly more contact interactions than non contact interactions. Bites are not distributed equally across the sites, however bites are not correlated with human or macaque density at a site (Spearman's rank r_s=.005). Across sites 382 (47%) of interactions involved tourist/taxi driver/tour guide provisioned food. However, each site varies substantially in the relative impact of food presence during interactions, and overall presence of food does not appear to have a direct relationship to interaction rates or aggression (Anderson *et al.*, in prep).

The focal follow data also demonstrate that contact interactions are significantly more frequent than non-contact interactions and that bites differ by site but not significantly and in a distribution different from that in the all occurrence dataset.

In the focal follow data adult male humans participated in more interactions than expected and adult female humans participated in less than expected. Adult male macaques are significantly overrepresented in interactions relative to their proportion in the population and adult female macaques are underrepresented in interactions relative to their proportion in the population. Immature macaques participated in interactions in proportion to their relative numbers (52% of interactions 54% of population).

Table 2. Summary of the focal follow data from four sites in Gibraltar.

	Apes' Den site	*Anglian Way/ St. Michael's Cave site*	*Cable Car site*	*Prince Philip's Arch site*	*Total: all sites combined*
Total # 15 minute focal follow samples	161	88	91	165	505
Total interactions	216	131	162	299	808
Interactions with contact	133	76	114	243	566
Interactions without contact	83	55	48	56	242
Total # bites	4	0	5	1	10

Table 2a. Statistical Analyses- All Occurrence sample interactions.

Site	*Interaction rate*	*Contact vs. non-contact intercations*	*Bites*
All sites Combined	Rates differ across sites (X^2=96, p<.001)	More contact than non-contact (X^2=129.9, p<.001)	Rates do not differ significantly across sites (X^2=6.8, p<.01>.001)
Adult Male Human participation	481/808 interactions - more than expected (X^2=26.8, p<.001)		
Adult Female Human participation	231/808 interactions - less than expected (X^2=46.8, p<.001)		
Adult Male Macaque participation	227/808 interactions - more than expected (X^2=74.45, p<.001)		
Adult Female Macaque participation	167/808 interactions - less than expected (X^2=23.24, p<.001)		

Table 3. Mean age/sex distribution of humans and macaques expressed as percentage of their total population during the observation period.

	Humans		*Macaques*
Adult male	47%	Adult male	16%
Adult female	42%	Adult female	30%
Child male	5%	Immature	54%
Child female	5%		

Focal data were also examined for the relationship between contact interactions, aggressive behaviour, and density of humans around the macaque at 0-1 metre intervals, 1-3 meter intervals, and at greater than 3 meter distances (within the limits of the site). No significant results were found (Anderson *et al.*, in prep). There does not appear to be a simple relationship between the number of humans in proximity to a macaque and the occurrence of contact interactions and/or aggressive interactions.

A comparison of the two main interaction sites of Apes' Den and Prince Philip's Arch were made to examine the participation of taxi drivers and tour guides in the interactions. These tourist couriers accounted for 18.1% of all recorded interactions observed during focal follows at Apes' Den and 33.8% percent of all interactions recorded during focal follows at Prince Philip's Arch. There are usually no more than a few (0-3) of these individuals at a given site during a sample period, and they make up a maximum of 7% of humans at Apes' Den and 9% of humans at Prince Philip's Arch at any given time. A simple chi-square test demonstrates that, at Apes' Den and Prince Philip's Arch, taxi drivers/tour guides are significantly overrepresented in interactions with the macaques. There are approximately 150 individuals in Gibraltar working in those roles at any given time, thus a substantial percentage of interactions occur repeatedly between these specific humans and a subsection of the macaques. If different taxi driver/tour guides are visiting specific sites and interacting with specific individuals then personality and historical factors may play important roles in explaining the variable results presented here.

Table 4. Participation of Taxi Drivers/Tour Guides at Apes' Den and Prince Philip's Arch.

Apes' Den 39/216 (18.1%) interactions, 7% of humans at site	$X^2 = 38.4$, $p<.001$ - more than expected
Prince Philip's Arch 101/299 (33.8%) interactions, 9% of humans at site	$X^2 = 202.8$, $p<.001$ - more than expected

Contextualizing the interactions

These data support an assertion that each of the sites differs from the others in their patterns of interaction. Apes' Den appears to be the best model for average patterns across the population, with Prince Philip's Arch having the highest rates of interactions and Cable Car Station the lowest. The differences in these interaction rates are likely attributable to differences in tourist flow and behaviour, behaviour of taxi driver/tour guides, and behavioural proclivities of individual macaques. Because this preliminary assessment suggests that simple measures of density and feeding do not predict interaction outcomes, a more detailed assessment at each site will be needed to identify the specific parameters underlying the variation in interaction patterns. Ongoing analyses of the focal follow data set focuses on the specifics of interactions, including patterns of variation within each macaque age/sex class with regards to various behaviours exhibited during the interactions and reaction patterns by the human interactors.

When interactions occur between humans and macaques, they are significantly more likely to include physical contact (both aggressive and non-aggressive contact) than no physical contact. These contacts range from hand to hand contact, to macaques climbing on the humans and sitting on their shoulders or head. During these very close interactions, humans' and macaques' faces can be in very close proximity, and both can exchange substantial flesh to flesh contact. Thus, there is a danger of potential pathogen transmission. However, given the recent analyses of pathogen load in the Gibraltar macaques (GONHS personal communication), the danger rests most largely on the threat of transmission from humans to macaques, especially of easily transmitted respiratory pathogens (aerosol dispersal via human exhaling). O'Leary and Fa (1993) report an instance of a viral pneumonia outbreak at Apes' Den in 1987 that killed all infants born in that year. Taken together, these sources of information suggest that contact interactions do pose a potential threat to the population of Barbary macaques in Gibraltar. Humans may also be at risk from exposure to the urine and faeces that can be easily transmitted through intimate contact via the hands and feet of the macaques.

There is not a simple relationship between human or macaque density and frequency of interactions. More humans at a site, more humans closer to a macaque, or more macaques at a site do not have statistically significant impacts on interactions in this dataset (see Anderson *et al.* in prep). This result appears counter intuitive, but may be due to the extreme habituation of the macaques to human presence and possibly the very high number of tourists present during summer months (the time period of this study) (Perez and Bensusan 2005). Anecdotal observations suggest that the macaques are quite accustomed to substantial human presence and may be fully acclimatised. Fa (1984) suggested that human presence and tourist provisioning may have a negative impact on macaque reproduction and lead to other physiologically stressful outcomes. It may be that the current provisioning strategy of feeding groups a total of over 43,000 kilograms of food annually (Perez and Bensusan 2005) lessens the impact of humans and that the multi-generational exposure to high numbers of humans in the current population of macaques has altered the perceptions and physiological responses of the macaques relative to earlier times (see Fa 1984 for demographic data 1936-1980).

Seasonal variation may also factor into the results presented here. Overall tourist density may play a role unnoticed in this temporally limited preliminary study. Perez and Bensusan (2005) report that June-August are the peak months for numbers of taxis and tourist driven cars entering the Upper Rock Nature Reserve suggesting that tourist numbers and provisioning opportunities are highest at this time of year. This overall tourist presence and provisioning may exceed a threshold wherein actual density or presence of food does increase interactions. This is supported by approximately 50 hours of observation (Fuentes, unpublished data, Kwiatt, unpublished data) during the months of January (2004 and 2005) and May (2005) where there appears to be a stronger relationship between presence of food and interactions (however, density and interactions remain unlinked as well in this small dataset).

Human adult males and macaque adult males are both overrepresented in interactions relative to their representation in the respective populations. Adult human females and adult female macaques are underrepresented. This deserves a much closer look and may be related to human cultural variables, the role of Taxi drivers/tour guides, and macaque sex differences in behaviour. There is also no simple relationship between the presence of food and interaction rates or aggression. Food from tourists or taxi drivers/tour guides does not appear to directly predict or always instigate interactions. Taxi drivers/Tour guides play a significant role in interactions at both Apes' Den and Prince Philip's Arch (at more than twice the frequency reported by O'Leary and Fa in 1991 for Apes' Den). They appear to be the main human interactors at these sites, at least during the peak tourist season. This suggests a systematic focus on their behaviour is needed.

The rate of biting by Barbary macaques in Gibraltar appears to be extremely low when compared with rates for Asian macaques (Fuentes and Gamerl 2005). The paucity of bites in this dataset makes any clear statement about casual relationships difficult. However, it is clear from these data that each of the sites has a different pattern of biting by macaques.

There are three other studies of human-macaque interactions that have produced some comparable data; Fuentes and Gamerl (2005), O'Leary and Fa (1993), and Wheatley and Harya Putra (1994). However, as the methodologies for the studies varied from the present one, certain direct comparisons are not possible. The only other similar study for this population was generated by O'Leary and Fa (1993) (see also Fa 1992). They examined 3128 Barbary macaque-human interactions at Apes' Den by macaque age/sex classes between July 2 and August 23, 1991. O'Leary and Fa report a maximum of 0.41 interactions per minute, whereas we report rates of 0.51 per minute overall and 0.58 per minute at Apes' Den. Our rates appear higher than the 1991 data set. They found that adult macaques were the age/sex most likely to interact with humans (34.7% of interactions), similar to our findings of over representation by adult male macaques (26% of interactions). They reported 33.3% of interactions involved adult females, which is higher than our overall 21% of interactions across sites. However, when examining just Apes' Den our results (25% of interactions by adult females) are closer to, but still distinct from, theirs. Their figure for immatures involvement in interactions, 32%, is substantially lower than our 51% (49.5% at Apes' Den) participation for non-adult macaques. This may be due to the very different population sizes and numbers of immatures in 1991 compared with 2004. O'Leary and Fa also report that adult male macaques were most likely to engage humans by grabbing food or exhibiting "begging" contact, and that food was involved in 61% of interactions (47% in our study). We did not have a clear correlation between food and contact. However, they report that infants and juvenile macaques were the age/sex class most commonly observed initiating contact with humans. This matches with our data, given that immatures make up the largest proportion of the population and were represented in interactions (in this data set) in proportion to their representation in the population. O'Leary and Fa also report that subadult male macaques initiated a disproportionately high level of aggressive attacks/threats. In our data set this also appears to emerge for one site only (Prince Philip's Arch) (Anderson *et al.* in prep).

Both O'Leary and Fa (1993) and Fuentes and Gamerl (2005) explain a lower aggression rate in adult female macaques by suggesting that because many of the adult females are carrying infants, human tourists may be favouring them as a provisioning target de-emphasizing the need for adult female macaques to be aggressive. However, this data set suggests that female Barbary macaques are under represented in all interactions (not just aggressive ones) relative to their proportion in the population, and thus there may be other explanations (Anderson *et al.* in prep). O'Leary and Fa also report that ~67% of interactions observed were contact interactions which falls very close to our 70% figure for all sites combined and 62% for Apes' Den. The mean numbers of humans per sample reported by O'Leary and Fa for Apes' Den in 1991 between 11:00 and 17:00 (> 50 humans per sample) falls slightly above our mean of 44.31 humans at Apes' Den. This is likely due to the increased number of macaque viewing sites in 2004 relative to 1991.

Wheatley and Harya Putra (1994) studying *M. fascicularis* at the Padangtegal and Sangeh sites on Bali found that feeding by tourists was significantly correlated with increased contact but not biting, and that the presence of food was significantly correlated with the total frequency of aggression by macaques towards humans. This differs from the findings in this study (see also Anderson *et al.* in prep). Fuentes and Gamerl (2005), for the same Padangtegal population of *M. fascicularis* also found a strong positive relationship between food, interactions and aggression. They report that adult male macaques were over represented in the interactions, especially in interactions that involved food and aggression. They also found that adult female macaques were underrepresented and that while immature macaques bit most frequently, adult and subadult males macaques bit adult human females most frequently. Fuentes and Gamerl report 48 bites in 420 interactions (11.4%) for focal follows whereas this study reports 10 in 808 interactions (1.2%) during focal follows (and 0.7% for the all occurrence follows (39 in 5489 interactions)). This tenfold difference in bite rate may be attributable to both the context in which the interactions occur and to species specific differences between *M. fascicularis* and *M. sylvanus*. The similarities between this study and Fuentes and Gamerl (2005) include: male over representation, female under representation and some differences in behaviour by macaques towards human age/sex classes.

Implications for management

It is clear that the situation in Gibraltar has changed since O'Leary and Fa conducted their study in the early 1990s. The diversification in population structure, the increase in numbers of tourists, and multiple interaction sites suggest that management priorities have to be at the same time, broad and specific. Broad in that there are general patterns of interactions that need to be managed. Specific in that each site appears to have a slightly different profile regarding interaction patterns and context.

In a broad sense, the data reported here suggest that taxi drivers/tour guides and tourist behaviour are important factors in driving the patterns of interactions with the macaques. Thus, managing the interaction patterns between the humans and macaques in Gibraltar will have to focus primarily on influencing human behaviour. While feeding is officially prohibited, no enforcement of the ban is enacted and a majority of the taxi drivers feed the macaques and encourage the tourists to do so also. This feeding can act to increase the chance of physical contact between humans and the macaques. Increased physical contact increases the health risk to both the macaques and humans. Because the Gibraltar macaques bite very infrequently they pose a minimal threat to humans in Gibraltar; however, human borne disease may be a significant threat to the macaques. GONHS currently provisions the macaques with sufficient food but the lure of high carbohydrate/protein "snacks" from taxi drivers and tourists will continue to provide an incentive for macaques to interact physically with humans. The data reported here suggest that food and human density are not direct correlates of interaction patterns, but that they are important elements in the interactions themselves. I suggest that increased management emphasis on reducing the amount of physical contact between humans and macaques should be a priority in order to minimise the risk to the macaque population.

In a specific sense, it appears that Prince Philip's Arch and Apes' Den sites receive the lions' share of interactions and thus may require the closest attention regarding enforcement of management practices regarding human-macaque interaction. However, because the macaque group structure and tourist flow patterns are relatively distinct at each of these sites slightly different approaches may be required.

Because the data presented here remain limited in temporal scope but illustrate the complexities inherent in the macaque-human relationship, I suggest that there is a need for the following research agenda regarding human-macaque interactions in Gibraltar:

1) Repeat the type of study presented here at other times of year: suggested times- Feb-March, October-Nov, Dec-Jan.
2) Conduct focal analyses of human taxi driver/tour guide behaviour at all sites using the same methodology.
3) Perform focal studies of specific macaque age/sex classes and individually known macaques at sites following the methods used here to build a database by individual for increased management potential and a better understanding of the macaque behaviour repertoire in a human interactive context.
4) Initiate and continue ethnographic and historical study of the care, maintenance, economic, and political aspects of the Barbary Ape's (see Shaw and Cortes this volume; Perez and Bensusan 2005).

Acknowledgements

This project was conducted with the sponsorship, collaboration, and assistance of Eric Shaw, John Cortes, Dale Laguea, Damien Holmes, Paul Rocca, Charles Perez and Keith Bensusan of the Gibraltar Ornithological and Natural History Society (GONHS), and Mark Pizarro (Gibraltar Veterinary Clinic). Project funding was provided by a University of Notre Dame Institute for Scholarship in the Liberal Arts Pilot Fund Program Grant. Additional support provided The University of Notre Dame Department of Anthropology and the Dean of the College of Arts and Letters. Special thanks go to the talented and dedicated 2004 University of Notre Dame Barbary Macaque Project Research Team: Meegan Anderson, John Canale, Tricia Hale, Noelle Easterday, Mary Eleazer,

Katie Hogan, Anne Kwiatt, and Steven Ludeke, and to the excellent macaque-human pathogen research team of Lisa Jones-Engel, Gregory Engel and Doug Cohn. I also wish to thank John Cortes and Keith Hodges for the invitation to participate in the Barbary Macaque CALPE conference and this book.

Bibliography

Aggimarangsee, N. (1992) Survey of semi-tame colonies of macaques in Thailand. *Natural History of Siam Bulletin* **40**: 103-166.

Anderson, M, Fuentes, A., Easterday, N., Eleazer, M., Hogan, K., and Kwiatt, A. (in prep) Patterns in aggressive interactions between humans and macaques in Gibraltar.

Engel, G.A., Jones-Engel, L., Suaryana, K.G., Arta Putra, I.G.A., Schilliaci, M.A., Fuentes, A., and Henkel, R. (2002) Human exposures to Herpes B seropositive macaques in Bali, Indonesia. *Emerging Infectious Diseases* **8** (8):789-795.

Fa, J.E. (1992) Visitor-directed aggression among the Gibraltar macaques. *Zoo Biology* **11**: 43-52.

Fa, J.E. (1984) Structure and dynamics of the Barbary macaque population in Gibraltar. In The Barbary Macaque: a case study in conservation. pp.263-306 J. Fa Ed. Plenum Press, New York.

Fa, J.E. and Lind, R. (1996) Population management and viability of the Gibraltar Barbary Macaques. In *Evolution and Ecology of Macaque Societies*. pp.235-262 J.E. Fa and D.J. Lindburg Eds. Cambridge Press.

Fuentes, A. (In press) Human culture and monkey behavior: Assessing the contexts of potential pathogen transmission between macaques and humans. American Journal of Primatology.

Fuentes, A. and Gamerl, S (2005) Disproportionate participation by age/sex class in aggressive interactions between long-tailed macaques (*Macaca fascicularis)* and human tourists at Padangtegal Monkey Forest, Bali, Indonesia. *American Journal of Primatology* 66(2).

Fuentes, A., Harya Putra, I.D.K., Suaryana, K.G.,Rompis, A., Artha Putra, I.G.A., Wandia, N., Soma, G., and Watiniasih, N.L. (2000) The Balinese Macaque Project: background and stage one field school report *Jurnal Primatologi Indonesia* **3**(1):29-34.

Fuentes, A., Southern, M., and Suaryana, K. (2005) Monkey forests and human landscApe's: is extensive sympatry sustainable for Homo sapiens and Macaca fascicularis on Bali. In Commensalism and Conflict: The primate-human interface. Patterson J Ed. American Society of Primatology Publications.

Fuentes, A. and Wolfe, L.D. (2002) Primates Face to face: the conservation implications of human-nonhuman primate interconnections. Cambridge University Press, Cambridge.

Huffman, M.A. (1996) Acquisition of Innovative Cultural behaviors in Nonhuman primates: a case study of stone handling, a socially transmitted behavior in Japanese macaques. In Social learning in Animals: the roots of culture. pp.267-289 C.M. Heyes and B.G. Galef eds., Academic Press.

Jones-Engel, L., Engel, G.A., Schillaci, M.A., Babo, R. and Froelich, J. (2001) Detection of antibodies to selected human pathogens among wild and pet macaques (*Macaca tonkeana*) in Sulawesi. *American Journal of Primatology* **54**:171-178.

Jones-Engel, L. Engel, G., Schillaci, M.A., Rompis, A.L.T., Putra, A., Suaryana, K., Fuentes, A., Beers, B., Hicks, H., White, R., and Allen, J.(2005) Primate to Human Retroviral Transmission in Asia. *Emerging Infectious Diseases* **11**(7):1028-1035.

O'Leary, H. and Fa, J.E. (1993) Effects of Tourists on Barbary Macaques at Gibraltar. *Folia Primatologia*, **61**:77-91.

Perez, C.E. and Bensusan, K.J. (2005) Upper Rock Nature Reserve: A management action plan. A Gibraltar Ornithological & Natural History Society Publication. Gibraltar.

Schlotternhausen, L. 2000. Town monkeys, country monkeys: a sociological comparison of a human commensal and wild group of bonnet macaques (*Macaca radiata*). Unpublished PhD. Dissertation City University of New York.

Wallis, J. and Lee, D. R. (1999) Primate conservation: the prevention of disease transmission. *International Journal of Primatology* **20**(6):803-825.

Wheatley, B.P. (1999) The Sacred Monkeys of Bali. Waveland Press, Inc.

Wheatley, B.P. and Harya Putra, D.K. (1994) Biting the hand that feeds you: monkeys and tourists in Balinese monkey forests *Tropical Biodiversity* **2**(2):317-327.

Wolfe ND, Switzer WM, Carr JK, Bhullar VB, Shanmugam V, Tamoufe U, Prosser AT, Torimiro JN, Wright A, Mpoudi-Ngole E, McCutchan FE, Birx DL, Folks TM, Burke DS, Heneine W. (2004) Naturally acquired simian retrovirus infections among Central African hunters. *Lancet*. **363**:932-37.

Zhao, Q. K. (2005) Tibetan macaques, visitors, and local people at Mt. Emei: Problems and countermeasures. In *Commensalism and Conflict: The Human – Primate Interface* edited by Paterson, American Society of Primatologists publication. Norman, OK Pp. 376-399.

The Gibraltar macaques: Origin, history and demography

E Shaw and J Cortes
The Gibraltar Ornithological and Natural History Society, Jews' Gate, Upper Rock Nature Reserve, P.O. Box 843, Gibraltar

Origins

The origins of the Barbary macaque population of Gibraltar have often been discussed, not least among the human population of the Rock, which shows a great interest in these inhabitants. The most often quoted popular sources of the animals are introduction by the Moors (who colonised southern Spain from the Gibraltar area from 711AD) and passage through a subterranean tunnel with an opening in St Michael's Cave. This clearly impossible source was considered plausible locally as recently as 1933 (Anon, 1933). These and other theories became part of the local folklore and appeared regularly in non-specialist publications aimed at tourists up to the 1980s (*e.g.* Garcia, 1975). The lack of fossil evidence of the species since the last glaciation virtually confirms the absence of the Barbary macaque as a naturally-occurring recent member of the fauna of Gibraltar.

Groves (this volume) lists a series of records of Barbary macaques in various parts of Europe in Carthaginian and Roman times, including Navan Fort, Armagh, Northern Ireland (Lynn, 1997) and Pompeii, Italy (Bailey *et al.*, 1999). As now, Barbary macaques no doubt attracted people with their behaviour, and it is likely that they accompanied human traders, settlers and conquerors around the Mediterranean much as other species of European mammals and birds were taken to distant regions of European Empires centuries later. Any incident resulting in an escape, or release of even a small number of macaques could easily have led to the establishment of a wild population, given the suitability of Mediterranean habitats and the adaptability of the species.

A wealth of non-European (mainly Arabic) texts exists as yet untranslated and therefore less accessible to European researchers, which refer to movements of north African peoples and migrations between north Africa and Europe, notably from Carthage (now Tunisia, from which the Barbary macaque disappeared only in the mid 19th Century) to Spain (M. Tlili, personal communication). These movements will in all likelihood have included pet animals. Phoenician remains have been found in Gibraltar and Carteia, a Carthaginian city, whose ruins are located approximately 6 km north-west of Gibraltar.

A reference to the Barbary macaque in the region of the Strait is contained in De Slane's (1965) translation of Abou Obeid el Bekri's work of 1068 AD. This refers to there being "no place on earth with more monkeys than Merça Mouça" (Jebel Musa, across the Strait from Gibraltar, where there is still a population of macaques (Mouna and Camperio Ciani, this volume)). Interestingly he makes reference to the monkeys imitating the actions "of men who pass nearby" stating that when they see men rowing in boats they grasp pieces of wood and copy them. Whether this bizarre report refers to the monkeys' habit of sometimes grooming in line or is a result of a knowledge at the time of the existence of macaques across the Straits in Gibraltar remains uncertain.

The first reference to the macaques in Gibraltar can be seen in Ayala (1782), who states:

> "There are the monkeys, who may be called the true owners, with possession from time immemorial, always tenacious of their dominion, living for the most part on the eastern side, in high and inaccessible chasms, neither the incursions of Moor, the Spaniards nor the English, nor the cannon nor the bomb of either, have been able to dislodge them."

Ayala, later quoted by several other authors, himself based his work on Portillo (ca 1620). Portillo does not mention monkeys in his in-depth treatise of natural history in Gibraltar, so the reference by Ayala cannot be taken as firm proof of their existence on the Rock before the British-Dutch capture in 1704. The Gibraltar Directory, an official publication of the 19th and early 20th Centrury, refers to "*a great number of these apes being sent into the Garrison in 1740*" (Samson, 1937). The source of this information is not known, and may be an assumption that "game from Barbary", allegedly brought to Gibraltar under the Governor's orders in the early 1700's, included monkeys. Clearly, though, by 1782, they were well established and well known.

History up to 1936

Anecdotal references to the Barbary macaques are regular in the Gibraltar Chronicle, which was first published in 1801. Some of these have been reported by earlier authors (*e.g.* Fa, 1991) and provide estimates of the numbers of macaques on the Rock varying from 3 to >130. Clearly these estimates are unreliable. While there is a suggestion that numbers were high in the late 1800's and then declined, they fluctuate widely and are not based on any reliable methodology. Many of the sources are letters to the Editor following casual observations of monkeys in the Town or recollections from years before. Others are reports following complaints. Several, like a series of articles by Lt.-Col. H.A. Sansom in 1937 did utilise official records ("*the official files kept in the office of the Colonial Secretary and that maintained by the Officer i/c Rock Apes*"). But the accuracy even of these must be in doubt. For example, A C Greenwood wrote in the Gibraltar Chronicle of 20th January 1937 that Sir Frederick Forrestier Walker, Governor between 1905 and 1910 had shown him the last three remaining "apes", "*all female*" and claimed that an attempt at introducing Moroccan males had failed as these had not been seen subsequently. However, writes Greenwood, "*one at least of these aliens must have found favour for, later on, the monkeys began to increase in number, and, in less than ten years have already lost their popularity by allowing "familiarity to breed contempt"*.

Furthermore, there were regular complaints of monkeys interfering with human residents during the 1920s when reported numbers rarely rose above 10. It is unlikely that such a small population would cause so much regular disturbance. These factors strongly suggest that not all animals present were recorded, either by casual observers or by the military personnel responsible for their feeding. Current experience of their behaviour by the management team of the Gibraltar Ornithological and Natural History Society (GONHS) confirms this likelihood. Only a small proportion of the animals known to be present are seen at any time, and individuals are known to "go missing" for extended periods, only to reappear. While it is clearly possible that monkeys were unofficially culled by shooting, the method used at the time, it is equally possible that they moved to other parts of the Upper Rock. This was more likely at that time than at present when the presence of other nearby groups restricts large shifts in these (Mehlman and Parkhill, 1988). The inaccessibility of most of the cliffs, together with the availability of natural food may have accounted for the apparent disappearance of the monkeys on different occasions. It must also be borne in mind that the Upper Rock was out of bounds to civilians and so the likelihood of casual observations of the monkeys would be low. The military personnel charged with feeding the monkeys at Apes' Den will not necessarily have sought them out in other locations. Undercounting of groups has occurred even in recent studies elsewhere (Mehlman and Parkhill, 1988). We suggest that undercounting in Gibraltar is likely to have occurred historically, especially when numbers were relatively low and when individuals were not clearly identifiable.

These are important points, as it has usually been assumed by researchers that anecdotal reports and military counts have been accurate. The drop of the population to four during the 1939-1945 World War and quoted by Fa (1981), must therefore be treated with caution.

One reference however needs to be highlighted, that is the report by Greenwood of very low numbers in the early years of the 20th Century, a point not recorded by Fa (1981). This led to the first report of attempted introductions. Greenwood records that "*one or two attempts had been made to introduce apes of the other sex* (*i.e.* males) *from the Moroccan coast*". This attempt at importation seems to be corroborated by H J King who wrote to the Gibraltar Chronicle in 1937 stating that "*between 1904-1905 there was imminent danger of the complete disappearance of the historical ape*", but noted "*the efforts of Col. C. Hill, who at different times between 1905-1906 imported two couples of Barbary Apes from Tangier*".

Several decades after this, there are reports of two young apes imported in August 1930 and another two in April 1931, all from Tetouan Province, and two in May 1932 and two in 1935, all four from Tangier.

History after 1936

One of the authors (ES) recently obtained the original hand-written record with details of the imports that were carried out between 1936 and 1946. These imports were co-ordinated through the British Consulate in Tetouan and consisted mostly of monkeys caught in the wild by trappers in the region of Chefchauoen, northern Morocco (W. Bruzon, personal communication) This details a total of 26 monkeys imported during that time when additionally there were 12 births and 25 deaths recorded. The recorded population in 1936 was a minimum of 5 animals, peaked in 1942 and following a catastrophic year with 11 deaths in 1943, required the importation of another 11 in 1944 to make the population in 1946 a minimum of 18 animals. Many of the imported animals were young (and therefore easier to trap), and records show that many did not survive for more than a few months. Bearing this in mind and the probability of under-recording, it is very likely that, contrary to popular belief, the population of macaques in Gibraltar did not fall as low as four during Second World War years. Considering too that births occurred during most of this period, and that these must be attributable to females of at least four years of age, it is likely that there still existed an unreported number of breeding animals somewhere on the Rock. The real contribution of the animals imported during the War years to the present Gibraltar population is therefore probably not as great as has generally been held. What is clear is that post 1930 imports do not account for all the genes in the current Gibraltar population. Indeed, the Middle Hill Group, believed to have been formed in 1946 (Fa and Lind 1996) could well have arisen from a nucleus of animals that had survived undetected.

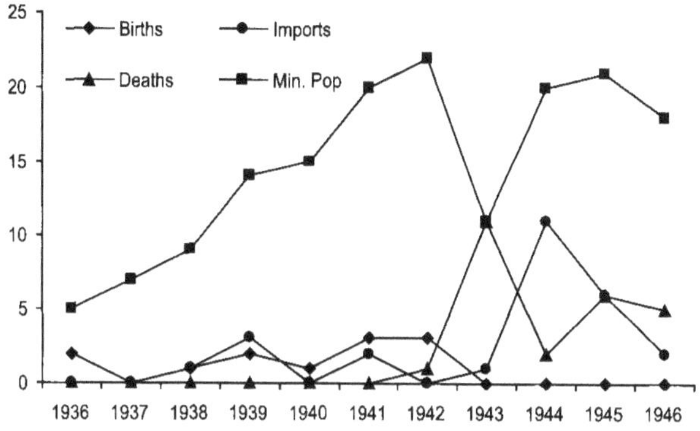

Figure 1. Recorded births, deaths and imports to the Gibraltar Barbary macaque population, 1936-1946.

Figure 1 traces the changes in the recorded Gibraltar population between 1936 and 1946. After 1946, under the control of the Army, the population was not allowed to increase, with an unknown number of animals being culled by shooting. Available records show that between 1968 and 1977, 13 animals were shot, 4 males and 9 females of ages ranging from 2 to 19.

Between 1949 and 1980, 75 apes were exported to zoos around the world, mainly in ones and twos. From then until 1998 when a group of 24 went to Daun, there were no further exports. Figure 2 shows the number of animals exported since 1940.

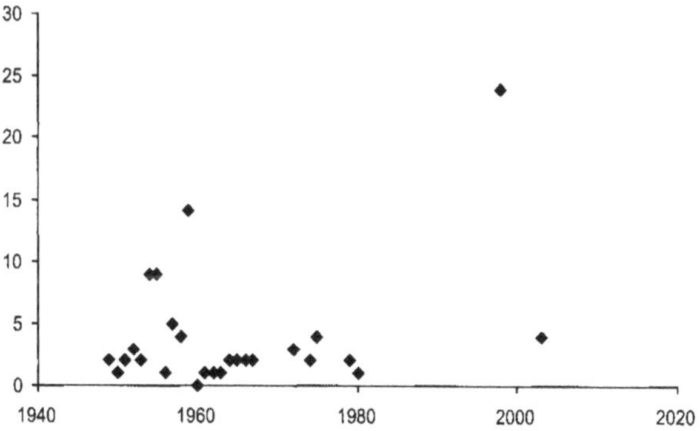

Figure 2. Gibraltar Barbary macaques exported since 1940.

The current situation

By 1946 at the latest, two distinct macaque troops were known to exist in Gibraltar, one at Princess Caroline's Battery (later Middle Hill) and the other at Apes' Den (= Queen's Road). The apes were contained in these two areas by culling by the military.

Figure 3 shows the fate of these groups up to 2003, while Figure 4 maps the present distribution.

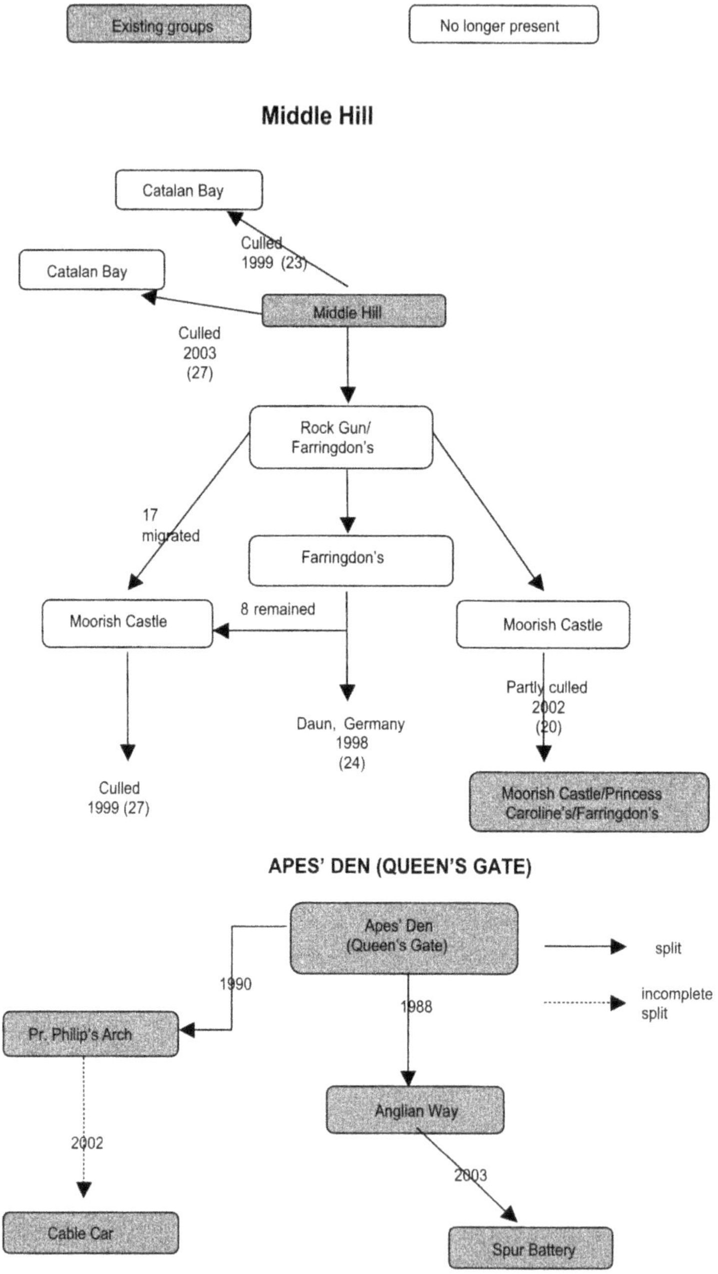

Figure 3. Fate of the Middle Hill and Apes' Den (Queen's Gate) troops up to 2003.

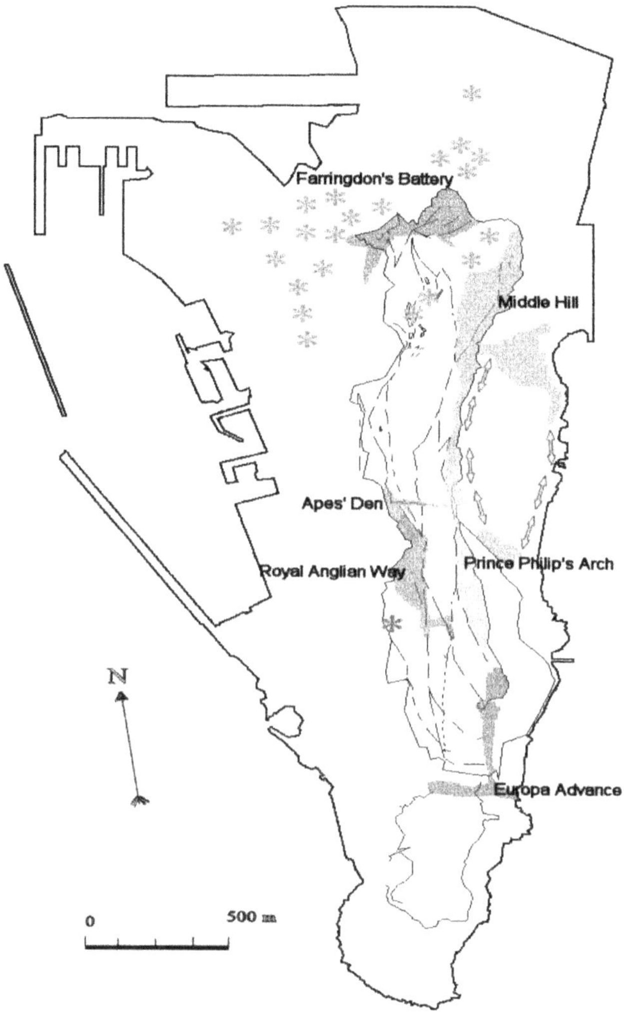

Figure 4. Present distribution of the Gibraltar macaque groups

Europa Advance is the former Spur Battery group split from Royal Anglian Way (= Royal Anglian Way II)
Prince Philip's Arch distribution includes Cable Car sub-group
Stars denote forays by the urbanised sub-group from Farringdon's

Since then there have been regular sub-groups of the Farringdon's group moving into the Town area and of the Middle Hill group into Catalan Bay. In addition, the Spur Battery splinter from Royal Anglian Way has now moved 1.5km south and is based around the refuse dump at Europa Advance Road and the surrounding cliffs. The Royal Anglian Way group has now spread to St Michael's Cave. All these extensions of home range, often preceding a split in the group, have occurred as expected towards areas where there were previously no macaque groups. These involve movements of up to 2 km for the Middle Hill group along the eastern side of the Rock, up to 1.5km for the Farringdon group into built-up areas, and of the Royal Anglian Way splinter group to Europa Advance. These movements are well within the ranges recorded in the wild (*e.g.* Mehlman 1989). The other groups are very sedentary by comparison.

Numbers of macaques on Gibraltar have been maintained at around the 200 mark in recent years, largely by the exportation or culling of groups. The main such events are recorded in Table 1, which shows a total of 24 monkeys exported and 50 culled between 1998 and 1999. This reduced the population overall, but even after the combined cull of 47 in 2002 and 2003, it has once again increased to nearly 250 (2006). Since 2003 three females have been exported to Portugal and there has been small-scale selective culling, usually of peripheral males or injured individuals in some of the groups, totalling no more than 20 animals.

Table 1. Numbers of animals in the various groups as at 1st January 1998, 2003 and 2006.

Site	*1998*	*2003*	*2006*
Middle Hill/Catalan Bay	62 *	65 ****	58
Prince Philip's	27	46	83
Anglian Way/Spur Battery/Europa Advance	28	36	49
Apes' Den	17	32	34
Farringdon	36 **	22	24
Rock Gun/Priness Caroline's	63 ***		
Total	233	201	248

* 23 culled at Catalan Bay, February 1999
** 24 exported, November 1998
*** 27 culled at Moorish Castle, October 1999, 20 culled at Moorish Castle, spring 2002
**** 27 culled at Catalan Bay, July 2003

Culling and exportation are by far the largest factors reducing the rate of increase of the population. Other causes of death recorded in recent years are shown in Figure 5. By far the most important is the running over by motor vehicles, to be expected in a Nature Reserve with excessive traffic flow (Perez and Bensusan, 2005). Anecdotal reports have referred to predation by "eagles" and on at least one occasion to scavenging of a dead animal by ravens (*Corvus corax*). However, Bonelli's Eagles *Aquila fasciata*, resident on the Rock until at least 1933 (Cortes *et al.* 1980) are capable of taking young macaques, and in 2005 there was evidence of one being taken by an Eagle Owl (*Bubo bubo)*, now resident on Gibraltar. However, these losses are well below those currently experienced by macaques in the Middle Atlas (Mouna and Camperio Ciani, this volume).

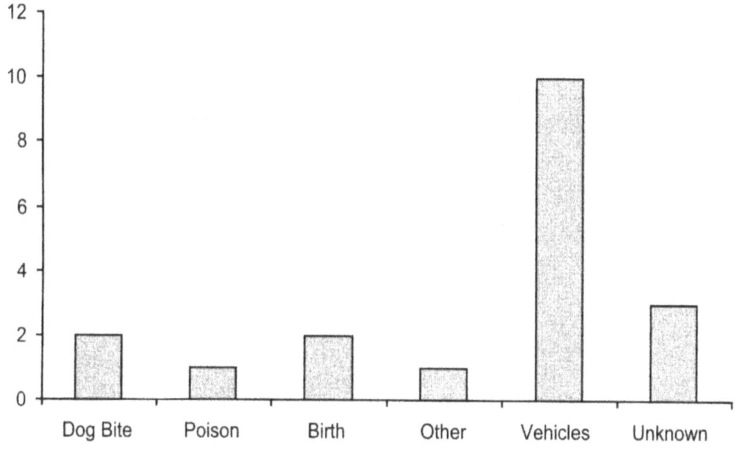

Figure 5. Causes of death of the Gibraltar macaques, excluding culling, 2000-2003.

Demography

The present population structure of the Barbary macaques of Gibraltar must reflect the many actions by Man over the past two hundred years, be they importation, culling, translocation or provisioning. Any demographical study must take these factors into consideration, and, with the possible exception of the Middle Hill troop up to 2003, demographical data must be viewed with caution. Bearing this in mind, we have used 2003 for a brief look at aspects of the demography of the Gibraltar macaques. Figure 6 shows the sex ratio in the different groups in 2003 while Figure 7 illustrates the sex ratios by age.

The overall sex ratio (male/female) is 0.99. The sex ratio of adults (over 5) is 0.66, and of immatures is 1.35. This excludes births in 2003. If these are added, the immature sex ratio is 1.10, somewhat closer to the 1:1 ratio of the adults. Work on wild macaques has revealed ratios generally lower than the Gibraltar overall figure ([e.g. 0.725 (Mehlman 1989) and 0.89 (Menard *et al.* 1990)]) where young "roaming" males have been removed. There are more young mature females than young mature males recorded, probably as a result of culling operations. There is overall a clear distribution bias in favour of young animals, and a paucity of older animals of both sexes, but in

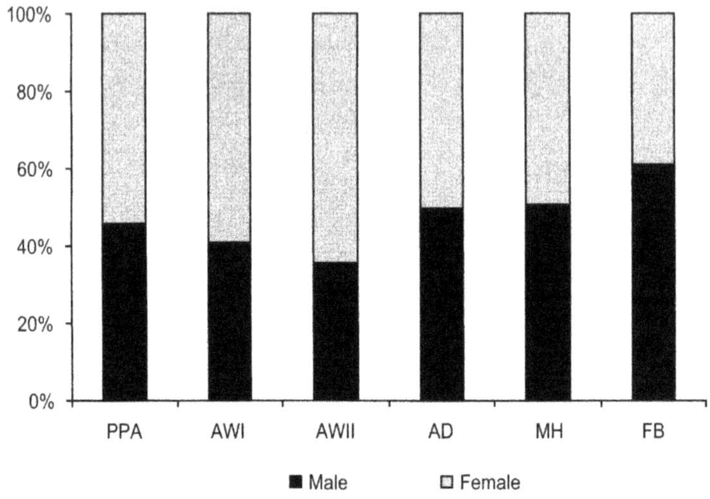

Figure 6. Gibraltar Barbary macaque sex ratios by groups, 2003

PPA = Prince Philip's Arch; AWI = Anglian Way; AWII = Anglian Way splinter group, later Spur Battery or Europa Advance Group; AD = Apes' Den (Queen's Gate); MH = Middle Hill; FB = Farringdon's Battery

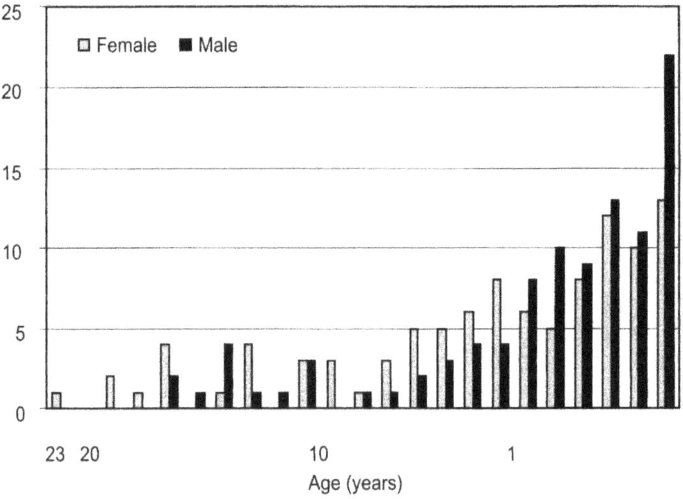

Figure 7. Gibraltar Barbary macaque sex ratios by age in 2003.

particular males. This again probably reflects not only infant survival, but also culling practices.

The overall adult/immature ratio is 0.77. For males it is 0.35 and for females it is 1.10, making immatures overall about 56.2% of the population. Mehlman (1989) records 46.9% immatures in his study, while Menard and Vallet (1996) recorded between 41% and 54%. Menard *et al.* (1990) gave the adult/immature ratio in rocky ridges in Algeria as 1.07-1.73 (mean 1.33). The difference is not surprising, since in the Algerian rocky ridges there was an indication of considerable infant mortality. In Gibraltar's rocky ridges the highest mortality factor is culling, not predation or starvation, and babies are not culled for sentimental reasons.

Discussion

Much has been speculated about the origins of the Gibraltar macaques. If a population of Barbary macaques became established following importation at any time before 1782 by any of a number of Mediterranean peoples, it is impossible to determine the likely source of animals. Possibilities range from northern Moroccan sources, including Jebel Musa and the Rif, if imported by the Moors, through to Tunisia (and possibly even Algeria) if they were distributed by the Carthaginians. Most recently, Modolo *et al.* (2005) and Modolo (this volume) used data derived from genetic studies to suggest possible introductions to Gibraltar, not just from Morocco, but also from Algeria. They found haplotypes in the Gibraltar population in common with Morocco (M02/M09 from the Rif; M02 from the Middle Atlas) and Akfadou in Algeria (M16). A Rif population in the area of Chefchaouen was also found to carry M16. Modolo *et al.* (2005) make several attempts at explaining the distribution of these haplotypes in relation to the Gibraltar macaques. They suggest that:

(a) it is likely that M02 came from Rif, not Middle Atlas animals;

(b) that surplus Gibraltar animals may have been exported to the Rif, explaining the occurrence of the Algerian M16 there.

We have seen above that importations from Morocco in historical times have been attributed to Tangiers and Tetouan. It is therefore likely that most of these animals originated in the Rif, although Middle Atlas animals can by no means be excluded.

There is no report of any export of Gibraltar animals to Morocco, and such exports are unlikely to have occurred. There is no record either of importation in recent centuries from Algeria, although this cannot be absolutely discounted as there have most probably been unrecorded imports at some stage. However, we suggest that most likely source of the Algerian genes in the Gibraltar macaques is in fact the Chefchaouen (Tetouan Province) population of the Rif from where there is historical and word of mouth evidence that animals were imported from the early 1900's to the 1940's, and that the presence of M16 there predates these imports.

Genetic and demographical studies on the Gibraltar macaques continue. Indeed demographic data are collected more systematically than ever before by the GONHS. Macaque management team (see Cortes and Shaw, this volume). The usefulness of such studies however is tempered by the history of the Gibraltar macaque colony and its intensive management by Man. We must therefore continue to delve into the past, following new sources of information, while taking the management of the population in the future in a direction that more closely represents natural processes. Research into wild macaques in North Africa is therefore not only of great relevance to the understanding of the Gibraltar population, but also to its future conservation and to its usefulness as a resource for primatological studies.

Acknowledgements

Dale Laguea and Damian Holmes are responsible for the keeping of the Bruce's Farm macaque database and for the compilation of data. The authors also wish to thank Bob Wheeler and the Gibraltar Garrison Library for assistance with historical information and William Bruzon for his personal account of the 1930's importation.

References

Anon (1933) *The Rock Apes: History and habits.* Gibraltar Chronicle article, 30 June 1933

Ayala IL de (1782) *Historia de Gibraltar.* Madrid. Sancha.

Bailey JF, Henneberg M, Colson IB, Ciarallo A, Hedges REM and Sykes B (1999) Monkey business in Pompeii – unique find of a juvenile Barbary macaque skeleton in Pompeii identified using osteology and ancient DNA techniques. *Mol. Bio. Evol.* **16** 1410-1414

Cortes JE, Finlayson JC, Garcia EFJ and Mosquera MAJ (1980) *The birds of Gibraltar.* Gibraltar Books. Gibraltar.

de Slane MG (1965) *Description de l'Afrique septentrionale par Abou Obeïd el-Bekri, traduite par Mac Guckin de Slane.* Edition revue et consignée. Libraire d'amerique et d'orient Adrien-Maisoneuve. Paris.

Fa JE (1981) Apes on the Rock. *Oryx* **16** 73-76

Fa JE (1991) The Rock Ape. Medambios. Gibraltar

Fa JE and Lind R (1996) Population management and viability of the Gibraltar Barbary macaques. In *Evolution and ecology of macaque societies.* Eds Fa JE and Lindburg DG. Cambridge University Press, Cambridge

Garcia J (1975) *The Famous Rock Apes of Gibraltar: A-Z Guide.* Medsun Publications. Gibraltar

Lynn CJ (1997) Excavations at Navan Fort 1961-71, County Armagh (original editor D. M. Waterman). *Northern Ireland Archaeological Monographs* No.3. Belfast: Stationery Office

Mehlman PT (1989) Comparative density, demography and ranging behaviour of Barbary macaques (*Macaca sylvanus*) in marginal and prime conifer habitats. *Int. J. Primatology* **10** 4269-292

Mehlman PT and Parkhill RS (1988) Intergroup ineractions in Wild Barbary Macaques (Macaca sylvanus), Ghomaran Rif Mountains, Morocco. *Am J Primatology* **15** 31-44

Ménard N and Vallet D (1996) Demography and ecology of Barbary macaques (Macaca sylvanus) living in two different habitats. In *Evolution and ecology of macaque societies* pp 106-131 Eds JE Fa and DG Lindburg. Cambridge University Press, Cambridge

Ménard N, Hecham R, Vallet D, Chikhi H and Gautier-Hion A (1990) Grouping patterns of a mountain population of Macaca sylvanus in Algeria-A fission-fusion system? *Folia Primatologica* **55** 166-175

Modolo L, Salzburger W and Martin RD (2005) Phylogeography of Barbary macaques (*Macaca sylvanus*) and the origin of the Gibraltar colony

Perez C and Bensusan K (2005) Upper Rock Nature Reserve: A Management and Action Plan. The Gibraltar Ornithological and Natural History Society. Gibraltar

Portillo AF (ca 1620) *Historia de la muy noble y mas leal ciudad de Gibraltar.* Manuscript.

Samson HA (1937) *The History of the Rock Apes.* In the Gibraltar Chronicle, 13th January 1937.

The Gibraltar macaques: Management and future

J Cortes and E Shaw
The Gibraltar Ornithological and Natural History Society, Jews' Gate, Upper Rock Nature Reserve, P.O. Box 843, Gibraltar

Introduction

The recent history of the Gibraltar macaques has been treated by several authors (Fa, 1981, Fa, 1991a, Shaw and Cortes, this volume). The present population consists of about 250 animals divided into six main groups with an additional one or two sub-groups, all based within the 97ha Upper Rock Nature Reserve. The climate of Gibraltar is typically Mediterranean, with hot dry summers and cool wet winters. No frost or snow is experienced. Figure 1 summarises climatological information for Gibraltar.

The colony has been established for an absolute minimum of 230 years, possibly much longer. The species, which occurred in Europe prior to the last glaciation, and is still found naturally in Jebel Musa 25km to the south-south-west across the Strait of Gibraltar, is clearly well suited to the Rock where it survived without official provisioning until 1918. A large population of at least 100 animals was established by the late 1800s (Shaw and Cortes, this volume), so that clearly there was enough naturally-occurring food to sustain considerable numbers. Allegations that they forayed into built up areas because of lack of food appeared in the Gibraltar Chronicle first in 1885. Like today, these incursions were in all probability natural movements, well within the ranging characteristics of wild macaques (*e.g.* Mehlman 1989), while the animals clearly took advantage, as they do today, of food offered by people.

The habitat occupied by the Gibraltar macaques has changed over the past 100 years or so, developing from open, grazed pseudosteppe and garigue to dense matorral scrub and developing woodland (Cortes, 1994). The open habitats of the 1700s and 1800s, described among others by Kelaart (1846) resembled open habitats now present in Jebel Musa (personal observation).

Subsequent vegetation regeneration, which took place during the 20th Century, will have reduced the potential foraging areas through the dense matorral stages, although some of these areas are now developing into woodland dominated by wild olive (*Olea europea)*, which is making the

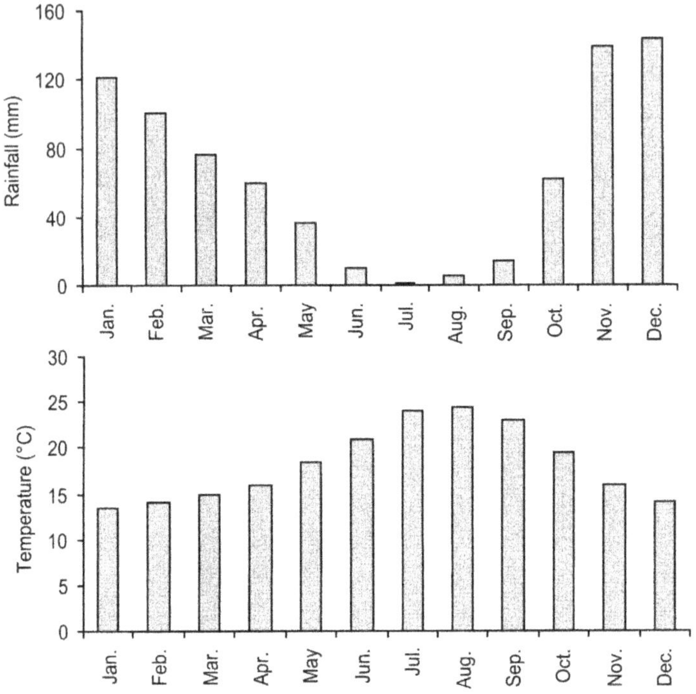

Figure 1. Climatological information for Gibraltar

ground accessible again. These changing habitats, together with the large areas of cliff that characterise Gibraltar have proved to be suitable for the population of macaques for centuries.

From 1918 to 1992, the responsibility for provisioning the macaques fell to the British Army, first a Master Gunner of the Royal Artillery, and later the Gibraltar Regiment – the so-called "Officer-in-charge-Apes". The decision to provision them was taken in the belief that contact with humans was due to lack of natural food, an assumption that has proved hard to dispel even to this day. From 1992, provisioning became the responsibility of the Gibraltar Government through the Gibraltar Tourism Agency, contracted first to Medambios, later handled directly and then, in 1999 contracted to the Gibraltar Ornithological and Natural History Society (GONHS), a Gibraltar-based non-Governmental Organisation.

The contractual obligations of GONHS include the laying out of food and water at established feeding sites and the cleaning of these sites. In addition, GONHS maintains a database of the population, co-ordinates and undertakes collaborative research with external institutions including universities and museums and performs other management tasks, including where possible removing monkeys from built-up areas in response to calls from members of the public and from the Police. In conjunction with the Gibraltar Veterinary

Clinic, which is also under contract to the Government of Gibraltar, the organisation regularly traps animals, which are tattooed, micro-chipped and photographed for individual recognition, subjected to a veterinary examination, tested for TB, campylobacter, salmonella and shigella, and vaccinated against rabies and distemper and treated for endo-parasites and other disorders (*e.g.*, skin disorders) where necessary. Occasional visits by specialist macaque veterinarians are also encouraged as part of the management programme.

The GONHS team collects a large amount of biometric data from each animal processed, which include body weight, total length, limb lengths and dentition. Blood and hair samples are taken and banked. In addition, all births and deaths are recorded and a database is kept of all the animals together with details of all known kin.

GONHS also takes on an advisory and advocacy role in relation to the macaques, being consulted by Government authorities, private companies and the Ministry of Defence on a wide variety of relevant matters. In 2004, following lobbying by GONHS, the responsibility for the management of the macaques was passed to the Gibraltar Government's Ministry for the Environment from the Ministry for Tourism.

Management practices

Control of numbers

Gibraltar Chronicle reports are quoted by Shaw and Cortes (this volume) with reference to records of population. These reports also make reference to the trouble caused by the "apes", since 1885, in built-up areas, gardens and orchards, to provisioning of the "apes" and to culling, the latter usually in response to such "trouble".

Historically, responses to the incursion of the monkeys in gardens, orchards, and parts of the Town, were various, and included the appointment of "watchmen", first referred to in 1888. Other steps included a Fortress Order (orders issued to the Garrison at the time) in 1910 forbidding the feeding of the apes by occupants of barracks and quarters, the passing of legislation in 1918 outlawing the feeding of apes other than at Queen's Gate (a law not revised until 2005), various instructions such as that to Police to prevent Hawkers from selling fruit at Southport Gate, and the increasing of the allowance of money to be used to purchase the food for provisioning the monkeys at Queen's Gate (later to be known as the Apes' Den) (*e.g.* from £21 per annum in 1918 to £48 per annum in 1921 "for a number not to exceed ten").

These measures apparently did not work. It is likely that unauthorised culling took place, and press reports show culling having been authorised as follows:

> 1888: culling allowed by special warrant from the Governor, numbers unknown;
>
> 1892: the Governor authorised 3 to be shot but also agreed that "numbers should be reduced".
>
> 1918: a suggestion that numbers be reduced to twelve.
>
> 1920: the Governor ordered that numbers be reduced to ten.
>
> 1921: numbers "unaccountably diminished" (see Shaw and Cortes, this volume).
>
> 1924: two large apes culled.

Throughout this time, and despite the occasional "agitation" against the monkeys, on the whole there was opposition to their being eliminated, expressed by the Governor, by both by the military and civilians, and even by the Secretary of State for the Colonies.

It is well known through personal communications that culling of the macaques by military officers continued into the 1980s. Reports are anecdotal, but it seems clear that most of the animals culled were adult or sub-adult males as well as other individuals that had become urbanised or that split from the established Apes' Den and Princess Caroline's Battery groups. A fairly large-scale exportation was carried out in 1998 and larger culls in 1999 and again in 2003 (Shaw and Cortes, this volume), all aimed at reducing the impact of urbanised monkeys.

Contraception is widely regarded as a humane alternative to culling, and some (less than 20) females in the Gibraltar population have been implanted recently as a trial. However, this method requires regular trapping of virtually all females of reproductive age, and may have behavioural implications. Consideration is being given to delaying first pregnancy and/or increasing inter-birth interval.

Roaming and urbanised monkeys

Most of the early reports of monkeys in human populated areas are in the Mount and Europa in the South District, and in the northern area of the Town. This suggests that two groups at least were responsible. A similar pattern is found today, with numerous incursions by successive sub-groups from the

Farringdon's Battery/Princess Caroline's areas in the north of the Town and less regular incursions of the Royal Anglian Way and Apes' Den groups in the South District. This is in addition to occasional males, singly or in small single sex groups, that roam further afield. Movement of macaques into the eastern side areas of Catalan Bay and Sandy Bay may have gone undetected in the 19th century when there were no tourists and the east side was sparsely populated by humans.

Current management practice includes attempts to chase and scare the monkeys away from built-up areas. However, they are resilient and once they have established that food is available, either in rubbish bin stores or at the hand of residents or visitors, they tend to increasingly focus their movements around these areas until they become fully urbanised, some individuals rarely venturing back into the Upper Rock. Removal of these food sources, such as the recent practice of caging in bin stores, will often result in the monkeys moving away, but only to other urban areas. The behaviour of the macaques is in many ways similar to that of habituated wild animals elsewhere, from bears to dingos (Rogers, 1993; Environmental Protection Agency, 2001; Creamer, 2005).

Although it is generally accepted that Barbary macaques move more at times of lower food availability, we consider the provisioning of the Gibraltar macaques to be more than adequate. Examples of home ranges and densities of wild macaques demonstrate that the distances moved in Gibraltar are well within what would be expected. For example, densities of macaques in natural habitat in Algeria include 28 animals per km^2 in Tigouate, and 13 animals per km^2 in Akfadou (Menard and Valet, 1996). The density on the Upper Rock is 102 per km^2. The Upper Rock Nature Reserve covers an area of only 97ha. This is larger than the home range of 39ha recorded by Menard *et al.* (1990), but considerably smaller than those reported by Mehlman (1989) which ranged from 300ha to 900ha, with an average of 720ha. The home ranges of the individual groups range from about 1 ha for Apes' Den to about 20ha for Middle Hill at the height of its population (65) when it was in the process of fission and frequented Catalan Bay. These ranges are small for the species and movement away from them cannot be unexpected.

Menard and Vallet (1996) found that their monkeys spent more time moving at the beginning of the dry season and in autumn. Whatever the true reason, which one would expect to be masked by provisioning, the wanderings of the Gibraltar macaques are consistently more marked at these times also.

The long-held, simplistic belief that the monkeys are roaming because they are hungry or not well fed not only denies the natural roaming characteristic of the species seen in the wild (*e.g.* Mehlman 1989) but also the fact that monkeys will prefer high calorific foods over natural vegetation or provisioned fruit and vegetables (Fa, 1991b).

The effect of tourists

The interaction between macaques and tourists in Gibraltar is well documented and the subject of several ongoing studies (Fa, 1992; O'leary and Fa, 1993; Fuentes *et al.* in press; Fuentes, this volume). Nearly 800,000 tourists visit the Upper Rock every year (Figure 2 and Perez and Bensusan 2005). Most of these are mainly attracted by the "apes", a total of about 250, of which not all are "on show" at any one time. The pressure on the individual monkeys is therefore tremendous, potentially in the region of 6000 tourists per monkey over a period of a year.

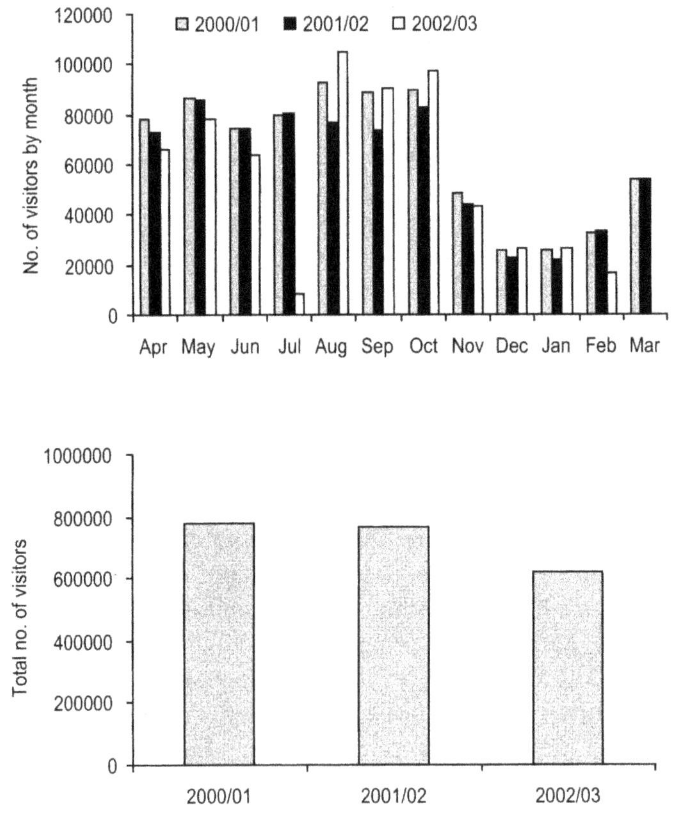

Figure 2. Tourist figures for the Upper Rock, Gibraltar

Despite laws and public campaigns, tourist handlers in Gibraltar continue to feed and handle monkeys regularly in Gibraltar (Fuentes, this volume), habituating them to expecting high calorie hand-outs. Apart from potentially negative effects on macaque health, this encourages the monkeys to harass tourists within the Upper Rock, leading to incidents of aggression, including biting (Fuentes, this volume), but also pre-disposing the macaques to approach humans when they venture into urban areas. Thus bag-carrying shoppers and school children are regularly approached by urbanised monkeys and their bags grasped. Familiarity with humans may also be the reason why urbanised monkeys regularly enter peoples' home. Such behaviour logically influences the authorities towards arriving at short-term solutions to the problem. As urbanised monkeys will not return to the Upper Rock, in the past such a solution has almost inevitably involved culling. The unregulated presence of tourists therefore interferes with management practices both directly and indirectly.

Whereas past studies have investigated macaque behaviour in the presence of tourists, current work by both Notre Dame University and Vienna University is looking into the behaviour of the tourists and their guides and how these are influenced by the behaviour of nearby macaques and other humans.

Provisioning

The macaques are provisioned on a daily basis by the GONHS team. Fresh fruit and vegetables, of a quality identical to that supplied to hotels and restaurants, are purchased regularly (Table 1). An amount equivalent to approximately 500g per animal per day is supplied, based on recommendations from Küster (personal communication) themselves backed by the study of Schemmelmann (1988) at Göttingen. The fruit and vegetables are cut into small pieces and distributed to the groups' feeding sites shortly after 0700hrs. every day. At the same time, fresh water is supplied and the sites cleaned of old food and faeces and disinfected. At around 1300hrs the sites are visited again, and grain in the form of wheat, as well as a seed mix including some sunflower seeds are scattered around their feeding areas.

Plans have existed since 1999 to improve the feeding sites by introducing ponds with circulating water and enlarging the area over which food can be scattered, to avoid clumping. This is difficult at present as many of the sites are on the sides of roads and spreading food on these can result in traffic accidents involving monkeys (see Shaw and Cortes, this volume). Budgetary constraints have delayed these improvements.

Despite occasional accusations that not enough food is provided, invariably food is left uneaten in all the sites and there is considerable consumption of natural foods. Fa (1991b) considers natural foods to be relatively insignificant in terms of weight and calorific value, although it is not clear that his data took full account of the considerable time that monkeys spend away from the traditional feeding points. Work is currently being undertaken to determine the true value of natural foods to the Gibraltar macaques.

Table 1. Breakdown of food provided to the Gibraltar macaques for the years 2000 and 2001 (excluding seeds).

Total Amount (kg /year)			
2000		*2001*	
Potato	9042.5	Potato	4512.5
Carrot	6305	Carrot	7280
Swede	1360	Red Cabbage	23
Orange	5639	Orange	2491
Cabbage	3901.5	Cabbage	5176.5
Onion	3627	Onion	3655
Sweet Potato	899.5	Sweet Potato	67.5
Pear	2607	Pear	2085
Celeriac	818	Cauliflower	486
Tomato	1552	Tomato	4829
Turnip	45	Pumpkin	6
Apple	5934.5	Apple	7467
Cucumber	1552.5	Cucumber	1322
Aubergine	10	Melon	3180
Banana	50	Celery	1300
Green Bean	20		
Total	43363.5	Total	43879.5

Discussion and recommendations

It is clear that, left unchecked, the Gibraltar population of Barbary macaques will continue to increase. As it does so, and certainly unless the behaviour of human visitors changes, they will come increasingly into contact with humans in urban areas leading to increased calls by the public for drastic action. This is precisely what led to the unfortunate and random culling in 2003 when a number of long-term research animals from Middle Hill were removed from Catalan Bay. Such measures must be avoided in the future.

It has been accepted in Gibraltar that the best form of drastic reduction in numbers is the removal of whole groups or sub-groups, either just after they have become urbanised (the culls of 1999 and 2003) or in the early stages when there are signs of potential splintering in groups adjacent to urban areas (the exportation to Daun, Germany in 1998). It is of great importance that possible destinations of macaque groups be identified in advance, so that pre-emptive measures can be taken. The Gibraltar authorities consider that the best option for reducing numbers is exportation, and consideration is currently being given to the moving of several groups to North Africa. Other less drastic measures suggested for future implementation is the culling of small numbers of carefully selected individuals on a regular basis. These would include sick or injured animals, or peripheral males, all of which would be more likely to fall prey to predators in the wild. As genetic information becomes available, decisions on individuals to be removed can be taken more sensitively with regard to the genetic health of the population.

The nuisance value of the macaques can also be reduced by proper management of the large number of tourists visiting the Upper Rock Nature Reserve. Proper enforcement of laws forbidding feeding of monkeys is required. Attempts in recent years to control this using volunteers have failed and it is increasingly clear that a full-time wardening system is needed, not just for the control of monkey feeding, but also for many other duties within the Nature Reserve (Perez and Bensusan 2005). It has been shown in Daun that the habituation of the macaques to rich food from humans is reversible (Küster, personal communication). This would greatly reduce macaque-human interactions on the Upper Rock and in urban areas, while tourists are quite happy to observe macaques feeding or grooming naturally without the need for jumping onto their shoulders or taking titbits from tour operators' pockets.

There are health implications also, both for tourists and for the macaques. While veterinary records show that the macaques are currently free of significant diseases, they are susceptible to human ailments. A false scare in 2003 that Herpes B was present in the Gibraltar macaque population led to near panic within the authorities as no structures were available to prevent regular macaque/tourist contact.

The Martin Plan (Martin, 1997) was presented to GONHS in 1997 and then transmitted to the Gibraltar Government, and formed the basis of recommendations made by Perez and Bensusan (2005). Those recommendations which are still valid are summarised in Table 2.

Conclusion

Since the mid 1990s, when the Anthropological Institute of the University of Zurich first made contact with GONHS, collaborative research with universities and other institutions, including the German Primate Centre, the University of Vienna, the University of Notre Dame, Indiana and Roehampton University, London has been undertaken regularly with the support of the Gibraltar based organisation. The results of such research are often useful in that they can be applied to the formulation of management practices. It is therefore important that researchers consider the applied implications of their work in submitting the results and conclusions of these to the managers. In parallel, the management team will strive to use good science and increasing practical experience to continue to improve on management techniques for the welfare of the Barbary macaques in their care and in the interest of the population and the species as a whole.

Table 2. Recommendations for future management of the Gibraltar macaques (based on Martin, 1997 and Perez and Bensusan, 2005)

a)	Macaques feeding sites should be improved, by providing large, covered areas for dispersal of food and suitable sources of water.
b)	Feeding should be twice a day, at 07:00hrs and 16:00hrs.
	This practice has not been fully adopted. The need to have staff on duty further hours was costed, but funds have not so far been made available. Instead, a second feeding with seeds and grain is carried out at 13:00hrs.
c)	Only natural foods should be used and should include native species such as figs *Ficus carica*, and the fruit of the strawberry trees *Arbutus unedo*, when in season.
	Native species are difficult to obtain commercially. The groups are provided with fresh fruit and vegetables and this has been augmented with grain.
d)	Two of the groups should be out of bounds to tourists and other visitors except for approved research purposes.
	The Middle Hill group remains largely isolated from the normal flow of tourists. No progress has been made in limiting access to other groups.
e)	One group should be accessible to visitors only on foot.
	The Anglian Way group's feeding site can be accessed only on foot. However, the proximity of a coach stop encourages the animals to spend much of their time on the roadside.
f)	Three groups should be visited in rotation only for one month at a time.
	GONHS recommended that visitors be allowed access to only one site for a month at a time. Proposal not adopted due to increased traffic congestion at the single site and likely resistance from the tour operators.
g)	GONHS should be entrusted, under contract, to establish an interpretation centre exclusively on macaque biology, to produce an up to date brochure on the macaques and to provide trained guides.
	Such an interpretation centre in one or two of the most visited sites (eg Queen's Gate) would be an extremely important and positive step, but has not yet been implemented.
h)	Steps should be taken to import a number of macaques from Morocco in order to introduce new genetic material.
	This should be done, given that recent studies have demonstrated that the population of macaques on the Rock is beginning to show the effects of inbreeding, although there remains a reasonably high level of genetic variability, showing that there is still time to counter its effect before it is to late.

Acknowledgements

Damian Holmes and Dale Laguea are responsible for keeping the Bruce´s Farm database on behalf of GONHS as well as being involved in the day to day management of the macaques. The authors also wish to thank Mark Pizarro and Rosinna Reyes of the Gibraltar Veterinary Clinic. The RAF Meteorological Office provided the weather data in Figure 1. Tourist statistics in Figure 2 were provided by the Gibraltar Tourist Board. Thanks also to Charles Perez for assistance with the figures.

References

Cortes JE (1994) The history of the vegetation of Gibraltar. *Almoraima* **11** 39-50

Creamer H (2005) Camping with Dingoes: visitor safety at a north coast campsite, New South Wales. National Conference of the Interpretation Association of Australia

Environmental Protection Agency (2001) Rick Assessment: Risk to humans posed by the dingo population on Fraser Island. Environmental Protection Agency, Queensland

Fa JE (1981) Apes on the Rock. *Oryx* **16** 73-76

Fa JE (1991a) *The Rock Ape.* Medambios. Gibraltar

Fa JE (1991b) Provisioning of Barbary macaques on the Rock of Gibraltar. In *Primate Responses to Environmental Change.* H.O. Box Eds Chapman and Hull. London

Fa JE (1992) Visitor-Directed Aggression Among the Gibraltar Macaques. *Zoo Biology* **11** 43-52

Fuentes A, Shaw E and Cortes J in press Humans, Monkeys, and the Rock: The anthropogenic ecology of the Barbary macaques in the Upper Rock Nature Reserve, Gibraltar. *Almoraima*

Kelaart EF (1846) *Flora Calpensis: contributions to the botany and topography of Gibraltar and its neighbourhood.* John Van Voorst. London

Martin RD (1997) *Outline proposal for effective management of the Gibraltar colony of Barbary macaques (Rock Apes).* Anthropological Institute, University of Zurich. Zurich

Mehlman PT (1989) Comparative density, demography and ranging behaviour of Barbary macaques (*Macaca sylvanus*) in marginal and prime conifer habitats. *Int. J. Primatology* **10** 4269-292

Ménard N and Vallet D (1996) Demography and ecology of Barbary macaques (*Macaca sylvanus*) living in two different habitats. In *Evolution and ecology of macaque societies* pp 106-131 Eds JE Fa and DG Lindburg. Cambridge University Press, Cambridge

Ménard N, Hecham R, Vallet D, Chikhi H and Gautier-Hion A (1990) Grouping patterns of a mountain population of *Macaca sylvanus* in Algeria-A fission-fusion system? *Folia Primatologica* **55** 166-175

O'Leary H and Fa JE (1993) Effects of Tourists on Barbary Macaques at Gibraltar. *Folia Primatol* **61** 77-91

Perez C and Bensusan K (2005) *Upper Rock Nature Reserve: A Management and Action Plan.* The Gibraltar Ornithological and Natural History Society. Gibraltar

Rogers L (1993) Studying habituated black bears In *Bears: majestic creatures of the wild* Ed Ian Stirling Rodale Press. Emmaus, Pennsylvania

Schemmelmann M (1988) Zur Ernährung Semifreilebender Adulter Berberaffen. Diploma thesis, Universität Göttingen

Bianca, a female Barbary macaque, with her 3 week-old baby in the deciduous oak forest of Akfadou, Algeria. Copyright D. Vallet

Living conditions and management of Barbary macaques (*Macaca sylvanus*) in European zoos and animal parks - status and potential for conservation of the species

J Küster
Herzberger Landstrasse 73, 37085 Göttingen, Germany

Introduction

The Barbary macaque is the only species of the genus *Macaca* occurring outside Asia. Its current habitats are high altitude (600 - > 2000 m) oak and cedar forests and rocky regions in Morocco and Algeria. The wild population is fragmented into several genetically isolated subpopulations and, due to habitat destruction and increasing human activity within their range, numbers are dwindling rapidly (Mouna and Camperio Ciani, this volume and references therein). IUCN lists the status of this species as vulnerable, and while measures need to be taken to stabilise wild populations, a fully integrated plan for conservation would also need to consider the use of captive animals.

Due to their adaptations to cold climate, Barbary macaques belong to the few primate species that can be kept outdoors in Europe throughout year. This makes them an attractive species for smaller zoos and animal parks that cannot afford expensive housing facilities. It also allows keeping the animals in large enclosures under "semi-wild" conditions. The ease of breeding, their generally docile nature and rich social behaviour add to their attractiveness.

This chapter surveys the living conditions and management of Barbary macaques in Europe. It focuses on basic information, *e.g.*, origin of founders, group size and composition, size and type of enclosure, diet, veterinary care, breeding and genetic management. The aim is to provide information to help optimise management in order to maintain a viable self-sustaining European population which can potentially contribute to future *in situ* conservation of this species.

Data set

A questionnaire was sent in September 2003 to nearly 60 zoos and animal parks which had been located from different sources, including an internet search of German facilities. Data from a total of 42 facilities were analysed: Of these, 26 were from facilities from nine countries that replied to the questionnaire or were known from the author's own visits (the "main data set") and 16 were from facilities that did not reply but from which limited information was obtained from other sources ("additional data set"). No information was available for ten facilities that did not reply. Eight facilities that had been contacted no longer kept Barbary macaques. The data set most probably does not reflect the true distribution of this species in the different European countries. Germany (19 facilities) is probably overrepresented (due to the internet search), while it is likely that in France, Spain and the UK more facilities exist than the nine which are considered here.

Results

Population size and origin of founders

The Barbary macaque is an established species in European zoos and animal parks. At 13 out of 31 facilities for which this information was available, the species has been kept for more than 20 or even 30 years. On the other hand, eight new facilities have been founded since 1998. Group size varies widely (Table 1). Half of the facilities have no more than 15 animals, but three large facilities in France (Kintzheim, Rocamadour) and Germany (Salem) keep between 100-300 animals, organised into several large social groups.

Table 1. Number of animals kept in the different facilities in 2003.

Animals	*Mds*	*Ads*	*Total*
1 - 5	2	1	3
6 - 10	5	4	9
11 - 15	7	2	9
16 - 20	4	2	6
21 - 25	3	3	6
26 - 30	1	1	2
31 - 35			
36 - 40	1		1
~ 100	1		1
~ 250	1		1
~ 300	1		1
Total	26	13	39

Mds, main data set: Ads, additional data set

Based on all facilities contacted, about 1300 – 1400 Barbary macaques currently live in captivity in Europe, nearly half of them in the three large enclosures in France and Germany (Table 2). About two thirds of these animals are of breeding age. The true size of the European population is however, certainly higher, since not all facilities holding Barbary macaques were contacted and animals are still imported illegally from Morocco as pets. For France alone, the number of these illegally imported animals is estimated to be close to 1000 (E. van Lavieren, personal communication).

Table 2. Animals kept in Europe in 2003.

	Mds	*Ads*	*Total*	*Breeding age*
Excl. KRS	328	~237	~ 565	~ 55 %
KRS	~ 650	—	~ 650	~ 75 %*
Total	~ 980	~237	~1215	~ 65 %
~ 10/facility for the rest	—	~130	~1345	

KRS: the 3 large enclosures at Kintzheim, Rocamadour and Salem
* A lower proportion of animals below breeding age in these facilities is due to contraceptive measures started in the 1980s.

While the origin of the founders of the captive population was not always known, Table 3 shows that the large majority of the animals are of Moroccan origin. The largest imports from the wild occurred in the 1970s for the three big enclosures in France and Germany (Kintzheim, Rocamadour, Salem). Only a very small number of these animals however, have subsequently been transferred to other facilities and hence their contribution to the gene pool of the remaining European population has so far been limited. Another import of 50 animals from Morocco also occurred in the 1970s for the then newly established facilities in Geiselwind, Rheine and Nuremberg. Today, descendants of these animals are found in more than 30 facilities. The largest importation of animals from Gibraltar occurred recently (1998) for the purpose of establishing a new facility in Daun, Germany.

Table 3. Origin of founder animals and number of sites their descendants lived in 2003.

	No. of animals	*No. of locations in 2003*
Wildcaught Morocco for KRS*	~ 250	3 (+1)
Wildcaught Morocco for Gw, Rh, Nb**	~ 50	> 30
Gibraltar	~ 30	2 (+ ?)
Wildcaught; confiscated/donated	~ 20	2
Zoos Morocco (some wildcaught)	~ 20	1
Total	~ 370	

* Kintzheim, Rocamadour, Salem
** Geiselwind, Rheine, Nuremberg

Living conditions, feeding and health measures

The size of enclosures for Barbary macaques in the facilities studied ranges from 30 m^2 to more than 25 ha. In eight facilities the animals are kept in cages with indoor and outdoor areas. Total floor area is 100 m^2 or less in all but one of these cases. In 17 facilities, the animals are living in outdoor enclosures ranging in size between 500 m^2 and 1 ha and equipped with "artificial" climbing devices (junglegyms, ropes, rocks). Enclosure size of 16 facilities is 1-25 ha with predominantly natural vegetation. Hence, in most facilities the animals are kept permanently outdoors with no adverse effects. Even at very low temperatures (-10 to -20°C), "shelters", if present, are rarely or never used. The most frequent size for enclosures is about 1 ha (9 facilities). Only eight of the 41 enclosures are larger. This size restriction has important consequences for group composition and breeding (details below).

At all facilities with enclosures larger than 0.5 ha, visitors are allowed to enter the enclosure and walk on a path through the monkeys' "home". With smaller enclosures visitors can watch the monkeys only from outside. Controlled feeding of unsweetened popcorn by visitors is allowed at four of 14 sites with visitor access. However, feeding of carrots, apples or peanuts and corn sold by the park is also allowed at three of 11 sites without visitor access. The behaviour of visitors (and monkeys) is monitored constantly by staff at seven facilities (three of which allow feeding). Irregular monitoring, especially on days with many visitors, is done at six further facilities (one of which allows feeding). Signs stating that feeding is prohibited are placed at two further sites, but monitoring is not carried out.

Fruit and vegetables are fed by staff at all 26 facilities comprising the main data set. At most (18) sites commercial monkey pellets are also offered. Small-sized grain with low energy content per item (wheat, rice, barley etc.) is fed at 15 sites. High-caloric seeds (sunflower, maize), nuts, bread/pastry as well as dairy products (curd, yoghurt) and animal products (canned dog food, mealworms) are offered at only a few facilities. No data on *how* food is offered (*i.e.* clumped or dispersed) are available.

Barbary macaques in the European population appear to be in reasonably good health. Five of the 26 facilities from the main data set have no routine health measures at all and at 17 facilities this is restricted to worming. Vitamins are routinely given at eight facilities. Vaccination against rabies is required by local authorities at some facilities where rabies is still present in the wild animal population. Tests for several diseases, either routine or when new animals were introduced, has shown that Barbary macaques are apparently less susceptible to tuberculosis than other macaques. To date, they have always tested negative for SIV and *Herpes simiae* making handling and close contact with people less risky than with other monkey species.

Breeding and genetic management

Barbary macaques in captivity breed regularly and in the large majority of facilities (29 out of 36) births occur every year. They are currently not breeding in five facilities because animals are either too young or only animals of one sex are present. Two institutions prevent breeding completely by reversible or irreversible measures (hormonal implants in females; castration of males). All but one of the facilities with intact animals of breeding age had births in 2003. A total of 95 infants was born in the 26 facilities of the main data set, 75 of which survived the most critical period of the first two weeks of life. Breeding is regulated in only a minority of facilities within the main data set, either by castration of all males (2 facilities) or hormonal implants in all or some females (7 facilities).

While the natural composition of Barbary macaque social groups is multi-male with an almost balanced sex ratio among adults (e.g. Paul and Küster, 1988), most facilities keep this species in harem groups (Table 4). Enclosure size is at least 1 ha when two adult males are kept. Groups with three or more males have enclosures of 5 ha and more. Several facilities with smaller enclosures (0.5 – 1 ha) attempted to keep two or three adult males but, due to the high level of male aggression during the mating season, were unsuccessful in the long run.

Table 4. Breeding group composition (mds and ads combined; only facilities with intact males and females of breeding age)

Number of males	*Mature males (≥ 4 yrs) (no. facilities)*	*Adult males (≥ 7 yrs) (no. facilities)*
0	1	1
1	9	20
2	2	1
3	2	2
4	2	0
≤ 10	1	2
≤ 15	1	0
16+	3	3
Total	21	29

Since Barbary macaque males reach sexual maturity at the age of four years, a new breeding male should be introduced into harem groups every 4-5 years to avoid paternal inbreeding (father-daughter incest). This was actually done in only two out of 17 facilities (in some, males have been exchanged, but more than 5 years ago). Unfortunately, since Barbary macaques react with great hostility and aggression toward newcomers, introduction of new animals is difficult. Although several facilities have tried to introduce new animals but failed, the replacement of a single male (preferably during the mating season) should be not that problematic.

For effective management, exchange of animals between sites, together with ability to individually recognise animals and maintenance of complete records of maternal and (as far as possible) paternal relationships are essential. Nevertheless, marking of *all* animals with a transponder chip and/or tattoo was reported in only half of the facilities of the main data set. Some animals are marked in five additional facilities and no animals are marked in seven. Most (19) facilities of the main data set have complete records of the maternal relationships of their animals, among them 12 of the 14 facilities with marked animals. However, complete knowledge of maternal relationships was also claimed by five facilities without marked animals.

Only three of 11 facilities with more than one male of breeding age did at least some paternity testing in order to keep track of paternal relationships. At two further facilities, material for later testing is collected routinely, while this aspect is neglected completely at the remaining six sites.

Discussion and conclusions

The present results show that living conditions of Barbary macaques in European zoos and animal parks vary widely. Group size ranged from less than 5 to more than 200 animals (wild groups: 18-88 animals; Ménard 2002), enclosure size from small cages of 30 - 50 m² to outdoor enclosures of more than 25 ha (home range size in the wild: 1-10 km²; Ménard 2002). Total population size in all facilities combined as well as number and origin of founder animals seem adequate for maintaining a viable European population of Barbary macaques in the future. However, an optimal management system which takes into account the conditions to which this species is adapted is necessary for this.

Feeding

Barbary macaques are largely vegetarian, opportunistic feeders. Their natural diet consists of flowers, fruits, seeds, leaves, roots and bark of a large variety of species of grass, shrubs, herbs and trees (Fa, 1984). Animal food (invertebrates) is only a very small part of their diet. Single food items are usually small, have a low energy content, and are distributed in a dispersed manner. Compared with other macaques, grass seeds are an important part of their diet (Ménard and Vallet, 1986). On the other hand, large fruits which form part of the diet of humans and other macaques (such as apples, oranges, grapes or bananas), do not actually occur in their natural habitat. Selection of food and feeding style should bear all these factors in mind.

Analysis of the questionnaires revealed that types of fruit and relative amounts of fruit and vegetables provided varies greatly between facilities. Low-caloric fruit, and small food items are recommended in order to avoid monopolisation, food-related aggression and excessive weight in the dominant animals. One piece of fruit for every ten pieces of vegetables seems a reasonable ratio. For similar reasons, high-caloric food like sunflower seeds, maize, nuts etc. should not be part of routine feeding, but restricted to situations with an increased energy demand e.g., temperatures below freezing. Bananas, although greatly loved, should be reserved for special situations such as the application of medicine. Bananas are monopolised completely by a few dominant animals and inevitably lead to increased aggression.

Milk products (curd, cottage cheese, yoghurt) are accepted by most animals, and can be used, especially if animals have no natural access to animal food. At a few facilities dog food or mealworms are also offered. This seems not to be essential, however, and many animals would not accept them at all.

Commercial monkey pellets are a convenient food offered at many facilities. They are easy to store, do not need preparation (cutting, washing), have a constant composition and contain all "necessary" minerals and vitamins. It should be kept in mind, however, that these pellets are usually developed for rhesus macaques whose natural diet is less frugal. It seems questionable whether pellets are advisable for captive animals that live outdoors and have additional access to a large variety of natural food items and Barbary macaques apparently stay healthy without the provisioning of pellets. Moreover, due to the moist climate in many parts of Europe, pellets will rapidly deteriorate when spread widely on the ground. They will therefore need to be placed in containers which will be monopolised with all the related problems of increased aggression and obesity.

A quantitative study of females at Wildpark Daun where the feeding schedule was changed weekly over a 2-month period revealed that aggression rate was twice as high under clumped feeding (4 food piles) compared with dispersed feeding (food scattered over an area of 500 m^2) regimes. Severe (contact) aggression increased almost sixfold when food was clumped (S. Roos-Kiefer, 2002 unpublished diploma thesis). Moreover, body weight of high ranking females increased to that approaching obesity (up to 16.5 kg), whereas body weight of low ranking females remained constant and within the normal range (10 - 12 kg). These results show that a feeding regime in which food is dispersed as much as possible is highly recommended. Not only the type of food but also a suboptimal feeding schedule can induce unnecessary aggression, stress, and related health problems (*e.g.*, obesity will shorten life span and can induce diabetes).

Contact between visitors and monkeys

Part of the attractiveness of outdoor enclosures is that visitors can enter them and come into close contact with the monkeys without barriers. If feeding is prohibited, no edible material should be allowed inside an enclosure. This should always be controlled at the entrance. Signs alone are clearly not sufficient, as many people are not willing or able to comply (personal observation). If feeding is allowed, only food which is only slightly or moderately attractive is acceptable in order to protect visitors from harassment by monkeys trying to steal it. Since Barbary macaques are strong and can become quite assertive in such situations, it is very important that visitors and monkeys are constantly monitored by dedicated staff. Moreover, it seems to be part of human nature to "experiment" in such situations without realising the potential danger of, for instance, "teasing" by preventing the animals from taking a food item, touching and above all, stroking. If visitors are not a potential food source, Barbary macaques ignore them more or less completely. Only food makes visitors attractive. Many years of working in enclosures with visitor access have revealed that all conflicts and accidents were ultimately related to feeding or touching. On the other hand, the author's own experience at Wildpark Daun where feeding is prohibited, has revealed that ordinary visitors without any scientific training also very much enjoy just observing the animals' natural behaviour.

If animals are kept in smaller enclosures where visitors stay outside, feeding, above all of high caloric food like peanuts, should be prevented. It will not only create health problems but also disrupt social behaviour resulting in, for example, constant begging and increased aggression among the animals about access to such food. It may also induce intense teasing of the monkeys by visitors who feel "safe" doing so with the barrier between the monkey and themselves.

Breeding and genetic management

Barbary macaques breed well in captivity and all but one of the facilities with intact animals of breeding age had births in 2003. Survival rates of 75 - 80 % will inevitably result in exponential growth rates with doubling of group size every 4 - 6 years. Hence, breeding can become increasingly problematic. Data on age specific rates of fertility and mortality from Affenberg Salem show that 60 % of all females *born* survive until the end of their 20-year reproductive period. During this time a female will give birth to 15 - 16 infants of which all but one or two will be successfully raised (Paul and Küster, 1988). Although many of these animals could be transferred to alternative locations in the past, it is almost certain that this will no longer be possible in the future. Hence, breeding regulation in order to reduce growth rates becomes a necessity for all facilities now. Several of the long-established facilities already regulate breeding (some have done so for 20 years). However, newly established facilities should not wait until the carrying capacity of their enclosures is reached, since then only a complete cessation of breeding can solve the problem of overpopulation.

Hormonal implants in females (Implanon®, Norplant® etc.) provide a reversible, safe and highly reliable method for regulating breeding in Barbary macaques. A single implant prevents pregnancies for 2-3 seasons, so that instead of a baby every year, interbirth intervals can be increased to 3 - 4 years. Most females with an implant remain sexually active and attractive to the males, although their strict temporal pattern of sexual behaviour is lost (Küster, unpublished observations). Longer interbirth intervals and fewer babies do not disrupt social behaviour. After removal of the implant, females become pregnant without undue delay, during the subsequent breeding season.

Since reversible methods for fertility regulation of males are still lacking, all available methods provide an "all or none" approach to fertility regulation. Barbary macaque females do not establish exclusive sexual relationships but have several sexual partners also during the period when conception occurs. Thus, castration of *some* males will not reduce birth rates, although the procedure has been carried out in a few facilities most probably to cope with male aggression. Even if reproduction needs to be completely prevented, as in those facilities that keep confiscated animals, permanent sterilisation of males or females has the drawback that animals are "lost" from the gene pool forever. Barbary macaque numbers in their native habitat, particularly in Morocco, are (still) rapidly decreasing. Census data of the largest population of this species (Middle Atlas region) revealed a drop from ca. 25000 to 4000 animals during the last 10 years alone (Mouna and Camperio Ciani, this volume). The total wild population is probably already below 10000, and there are no indications that this trend will be reversed in the foreseeable future. Thus, it is quite possible that Barbary macaques in European zoos and animal parks may soon become an extremely important part of the overall gene pool for the species as the viability of the wild population becomes increasingly uncertain.

Barbary macaques live in multi-male multi-female groups with a balanced sex ratio among animals of breeding age. Nevertheless, at almost all facilities surveyed (except the few very large enclosures), only one group with a single breeding male was kept. Undoubtedly, high levels of aggression during the mating season among adult males make it difficult to keep more than a single adult male in a restricted space. It seems, however, that the 1 ha size of enclosures represents the limit for an area that a single male can monopolise. Thus, an increase in enclosure size above 1 ha may facilitate the keeping of additional males. This is not only desirable in relation to genetic management but is urgently needed to prevent the creation of high numbers of "surplus" males. Moreover, adult Barbary macaque males have an elaborate system of support and alliance formation. If males intervene during male-male conflicts, they almost always support the lower ranking victim (Küster, unpublished data). This pattern effectively minimises power differences between the opponents, increases the risk of escalating conflicts for the dominant males and greatly reduces open aggression and wounding. Hence, it *might* be possible that it is easier to keep 5 - 6 adult males (albeit of different ages) with their flexible support pattern in a 1 ha enclosure than only two or three.

Unfortunately introducing new animals in established groups is difficult in Barbary macaques, since they react with extreme hostility and aggression towards newcomers. Several facilities have tried to add new animals but failed. Exchanging single males among harem groups for genetic management should not be that problematic, however. Other "tricks", such the exchange of male newborns between females, as routinely done in some rhesus breeding colonies, are probably not feasible between different sites. More information about this topic needs to be exchanged between the facilities.

While harem groups facilitate keeping track of paternity, since no tests are necessary, paternal inbreeding becomes a problem within short periods of time. Barbary macaques strongly avoid sexual relations with maternal kin (at least as long as there are alternatives), but they do not avoid sex (and reproduction) with paternal kin (Küster, Paul and Arnemann, 1994). Since paternity is fairly distributed among the males in multi-male groups, each additional male will decrease the chance of inbreeding proportionally. For this reason it would also be desirable to keep at least some young adolescent males (4-5 year olds) in harem groups. They do already reproduce (Küster, Paul and Arnemann, 1995), but do not yet evoke high rates of aggression in adult males. It should be stressed that the effects of inbreeding are not always apparent immediately. The fact that seemingly viable offspring result from some particular father-daughter inbreeding does not indicate that the species is immune to its potential negative consequences. Small sample size just makes it more difficult to detect these consequences. It seems much more desirable to prevent inbreeding *before* the consequences become suddenly apparent for everybody, and elaborate breeding programmes become necessary to save the species.

If there is more than one potential father for an infant, paternity determination becomes necessary. Large scale paternity testing in the population of Affenberg Salem has revealed (among other things) that males of all ages have the chance to reproduce and that observational data are not reliable for determining paternity. Males with a low copulation frequency may nevertheless have a high reproductive success and *vice versa* (Küster *et al.*, 1995). Today methods for testing paternity (DNA fingerprinting) are well established, and although they may not yet be routinely available, storage of samples now in order to carry out such tests in the future should be encouraged. A few hairs are sufficient and can easily be stored in a freezer for years and if this is done whenever an animal is handled for other reasons, collection of these samples requires little additional work.

Effective management relies on individual knowledge of the animals and their maternal kinship relations which influence social structure and interaction patterns in a fundamental way (for a review see Paul, this volume). Nevertheless, according to this study, marking of animals by tattoo or chip is carried out in only half of the facilities. Although it is perfectly possible for an experienced observer to recognise Barbary macaques individually without checking a tattoo or chip, this is a "high risk" strategy, which relies on the constant, uninterrupted presence of such a person. Any confusion of individuals after a gap in observation will result in the loss of information on maternal relationships, which is difficult to reconstruct later. Marking animals may be time-consuming especially in large outdoor enclosures where animals have to be trapped, but there is no satisfactory alternative.

The total number of founder animals identified in this study provides a promising situation for maintaining a viable population of Barbary macaques in the future. However, in order to reduce the risk that every facility will become a genetic "dead end" for the respective individuals, improved exchange of animals between different facilities is needed. Animals from *all* sources (including confiscated pets after rehabilitation as is done, for example, at the Sanctuary for Exotic Animals "AAP" in the Netherlands), should be considered. It would be a pity if actions were not taken until the point is reached when the captive population of Barbary macaques also becomes threatened due to genetic impoverishment.

Acknowledgements

I greatly acknowledge the help and cooperation of the many people of the different zoos and animal parks who made the compilation for this report possible. Special thanks go to Achim Johann, Director of Naturzoo Rheine, who provided valuable information on many additional facilities.

References

Fa JE (1984) Appendix II: Diet of the Barbary macaque in the wild. In *The Barbary macaque: A Case Study in Conservation* pp 347-355 Ed JE Fa. Plenum, New York

Küster J, Paul A and Arnemann J (1994) Kinship, familiarity and mating avoidance in Barbary macaques, *Macaca sylvanus Animal Behaviour* **48** 1183-1194

Küster J, Paul A and Arnemann J (1995) Age-related and individual differences of reproductive success in male and female Barbary macaques (*Macaca sylvanus*) *Primates* **36** 461-476

Ménard N (2002) Ecological plasticity of Barbary macaques (*Macaca sylvanus*) *Evolutionary Anthropology* Suppl. 1: 95-100

Ménard N and Vallet D (1986) Le régime alimentaire de *Macaca sylvanus* dans differents habitats d'Algerie: II. Régime en forest sempervirente et sur les sommet rocheux [The diet of *Macaca sylvanus* in different habitats of Algeria: II. Diet in evergreen forests and on rocky summits] *Revue Ecologique* (*Terre Vie*) **41** 173-192

Paul A and Küster J (1988) Life-history patterns of Barbary macaques (*Macaca sylvanus*) at Affenberg Salem. In *Ecology and Behavior of Food-Enhanced Primate Groups* pp 199-228 Eds JE Fa and CH Southwick. Alan R Liss, New York

Roos-Kiefer S (2002) Der Einfluß der Nahrungsverteilung auf das Konkurrenzverhalten weiblicher Berberaffen (*Macaca sylvanus*): Eine experimentelle Studie im Freigehege des Wildparks Daun [Influence of food distribution on competitive behaviour of female Barbary macaques (*Macaca sylvanus*): An experimental study at Wildpark Daun]. unpublished Diploma thesis, University of Bochum/Germany

Conservation genetics of Barbary macaques

L Modolo
Anthropologisches Institut und Museum, Universität Zürich, Winterthurerstrasse 190, 8057 Zürich, Switzerland

Introduction

Habitat destruction and other human disturbances are threatening many species with extinction. In addition to the direct threat that these factors pose, species with reduced population size have higher rates of inbreeding and suffer from a loss of genetic diversity. Inbreeding leads to reduction in reproductive fitness in naturally breeding species in the wild (Saccheri *et al.*, 1998), as well as to the accumulation of deleterious alleles due to genetic drift, making small and isolated populations vulnerable to extinction (Simmons and Crow, 1977). Inbreeding is also a major risk factor for *ex situ* populations of threatened species, as it is known to elevate extinction risks in captive environments (Frankham, 1995a). Thus, since population viability and decline are influenced not only by ecological factors, but also by genetic ones, molecular markers can be used as an imaginative and informative way of measuring biodiversity (Beebee and Rowe, 2004).

A major interest in conservation genetics is maintenance of genetic variation in endangered or threatened species and continuation of gene flow between isolated populations. The study of genetic structure of small subpopulations can enhance our understanding of the genetic consequences of isolation both in the wild and in captivity, which is of central importance for conservation biology. Furthermore, it can supply vitally important data needed to develop plans for effective genetic management of captive populations and, in the long term, for improved conservation management of remaining wild but similarly isolated populations.

Barbary macaques were formerly widely distributed throughout North Africa, but their distribution is now confined to just a few forested areas of Morocco and Algeria (Figure 1; Taub, 1977; Fa 1984a). Progressive destruction of existing habitats by deforestation has led to extreme population fragmentation (Taub, 1977). Effects of isolation and fragmentation are not confined to natural populations; semi-captive colonies also suffer from restricted gene flow as a consequence of isolation. In addition to the natural

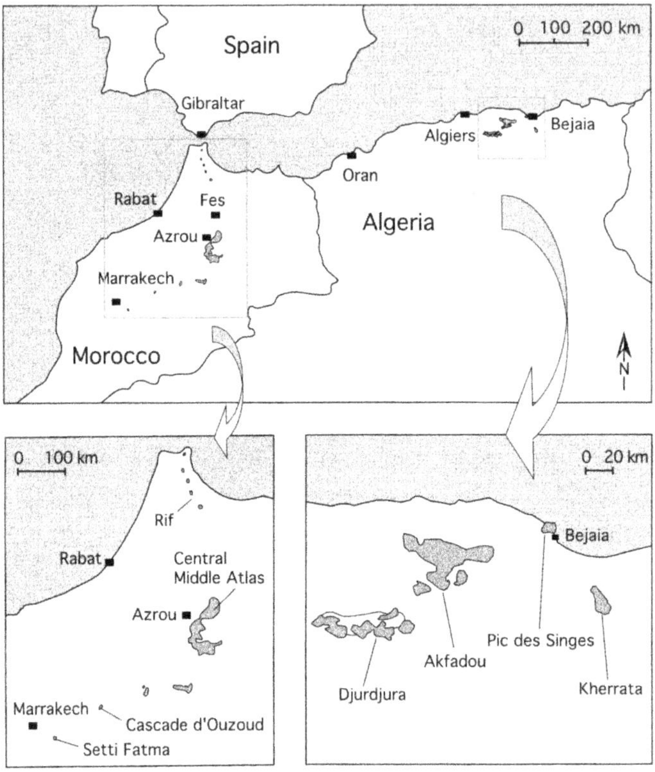

Figure 1. Map of Morocco and Algeria showing approximate areas of distribution of *Macaca sylvanus* (according to Taub *et al.*, 1977; Fa, 1984; von Segesser *et al.*, 1999). For Algeria, four of the seven remaining wild populations, from which samples were obtained, are highlighted. In Morocco, there are three main areas of distribution: Rif, Middle Atlas and High Atlas; the latter is indicated as Cascade d'Ouzoud and Setti Fatma.

populations of Barbary macaques, several artificially maintained colonies also exist, such as the free-ranging colony of Gibraltar. The provisioned Gibraltar population is particularly subject to genetic side-effects because of limited space and complete isolation from other subpopulations. In addition to the Gibraltar macaques, there are 3 other free-ranging captive colonies in France (Kintzheim and Rocamadour) and Germany (Salem). The colonies, each containing about 300 individuals, are derived from 260 founders imported from the Middle Atlas in Morocco. Additional groups of 10-50 animals deriving from various sources are found in smaller outdoor enclosures or zoos. A summary of the world Barbary macaque populations is provided in Figure 2.

Barbary macaques in the wild

Barbary macaques used to be widely distributed in the humid zone of the once entirely forested Maghreb, inhabiting diverse environments ranging from

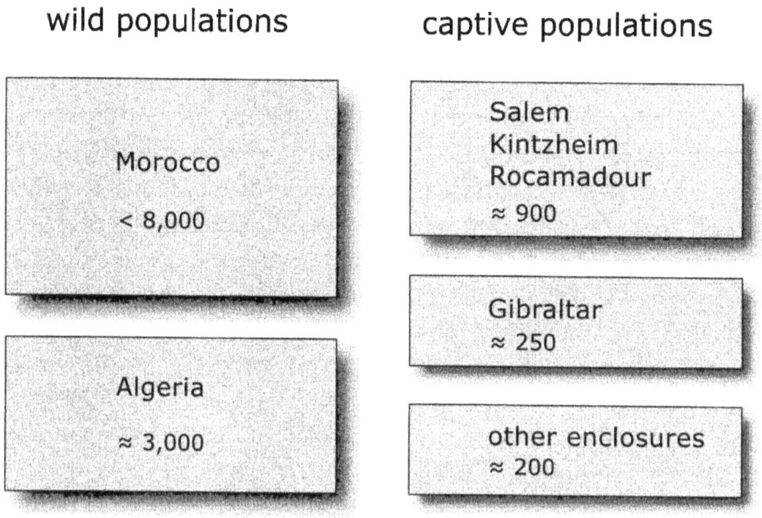

Figure 2. Breakdown of the global population of Barbary macaques (*Macaca sylvanus*), both wild and captive (after von Segesser, 1999); numbers modified following Camperio Ciani (2003) Estimates of captive populations can vary considerably due to fluctuations in reproduction and management strategies.

lowland shrub to high-altitude coniferous forests between Morocco and Tunisia (Taub, 1977). For 4000 years, the Maghreb has been continuously subjected to human influence by successive civilizations following Roman and Arab invasions. Increasing human disturbance and the resulting disappearance of forests has greatly affected the Barbary macaque population. Two decades ago, the maximum population size of *M. sylvanus* in the wild was estimated at 20,000 individuals located in a few refuges in Morocco and Algeria (Taub, 1977, 1982). Since then, the wild population has apparently decreased to 10,000 (Camperio Ciani, 2003), with habitat loss being regarded as the principle factor responsible for this decline. (Fa, 1984a; Taub, 1977; for a summary of reports see Taub, 1982). Remaining wild subpopulations are completely isolated in forest fragments, separated by extensive intervening built-up areas, such that natural genetic exchange between them is completely ruled out. In Morocco in particular, the population decline is attributed to the extensive forest destruction due to overgrazing by sheep and goats (Camperio Ciani *et al.*, 2005; this volume) and the resulting increased occurrence of degraded underbrush. The density of macaques is declining even faster than the forest itself.

In a recent study, phylogeographic relationships among surviving wild subpopulations of Barbary macaques in Algeria and Moroccan were examined by sequencing a segment of the most variable part of mitochondrial DNA (mtDNA) (Modolo *et al.*, 2005). Twenty-four haplotypes were identified, revealing significant genetic structuring among Barbary macaques (Figure

3). Haplotypes of the different lineages were clustered according to geographical relationship. Accepting an origin of the genus *Macaca* about 5.5 mya (million years ago) (Delson, 1980; Tosi *et al.*, 2003) an initial split between Algerian and the majority of Moroccan populations was estimated to have taken place about 1.6 mya, whereas a subsequent separation between lineages for Morocco and Kherrata (Algeria) was dated around 1.4 mya (Table 1). The deep genetic structuring among Algerian populations probably arose through extinction of intermediate populations and physical barriers to dispersal. In species with female philopatry, mtDNA variation within groups is very low, whereas interpopulational differences can be large (Melnick and Hoelzer, 1992). Barbary macaques are characterised by male-biased dispersal and female philopatry; hence groups are composed of female kin and immigrant males. The pronounced philopatric nature of female macaques has been shown to preserve multiple mtDNA types and can lead to local fixation of deeply divergent lineages (Hayasaka *et al.*, 1996; Tosi *et al.*, 2003). Despite female philopatry, the distances found between the three Algerian lineages of Barbary macaques were still surprisingly large. Interestingly, the single social group in Pic des Singes contains two completely distinct sublineages (M17/M18; Figure 3) differing by 23 mutational steps (4.9%). This finding could indicate a recent isolation of Pic des Singes from previously connected populations, because the two haplotypes are affiliated to neighbouring populations. A similar distance was found between the populations of Djurdjura and Akfadou, consistent with their close geographic proximity (50km distant). This suggests that dispersal might have been possible to some extent at a time when intermediate populations and forest corridors were still present. Despite this small geographical distance, however, these two populations have markedly distinct haplotypes and their divergence time is estimated to be around 0.9 million years ago. It is therefore possible that the present habitats do not correspond to the previous distribution or that male dispersal was restricted to a small area, keeping local groups isolated from other groups. This would further suggest that the observed deep differences in mtDNA are more likely to be due to historical isolation than to recent fragmentation.

Conversely, Moroccan haplotypes from three geographically distinct populations were grouped into a tight monophyletic cluster, revealing a very close relationship among them. This is an interesting case, because geographical distances between Moroccan populations are much larger than between the distinct Algerian lineages. The apparently oldest haplotype (M01; Figure 3), from which the central and most abundant haplotype (M02) derives, is exclusively located in the Middle Atlas. Because the two other Moroccan populations, Rif and High Atlas, were found to possess haplotypes deriving from the central haplotype and occurring only in these two regions, this central

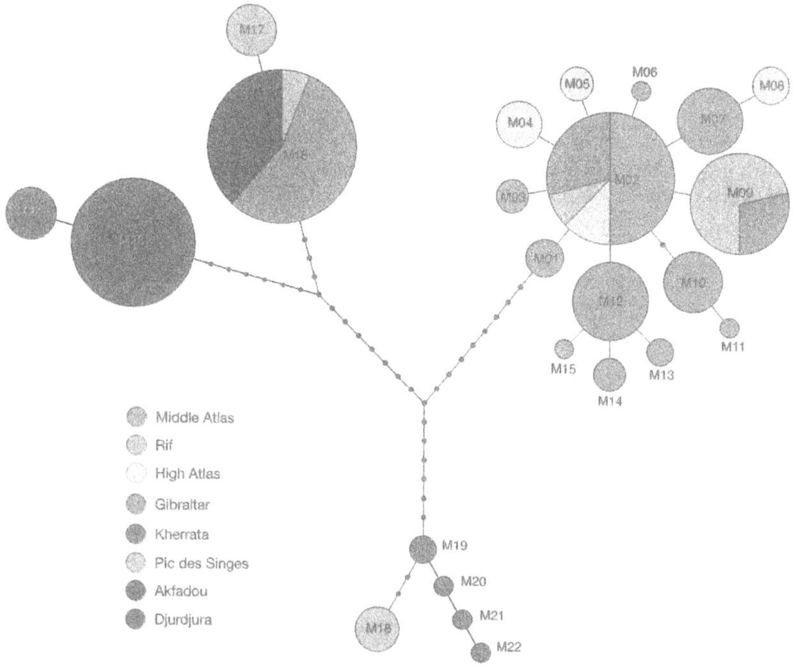

Figure 3. mtDNA haplotype network of Barbary macaques generated by Modolo *et al.* (2005). Each line between points represents a single mutational step. A haplotype is represented by a circle whose size is proportional to the number of individuals showing that haplotype. Haplotypes are colored to match the respective population in the legend. Distribution of the present natural populations and the Gibraltar colony is illustrated in Figure 1. Reproduced from Modolo *et al.*, 2005, with permission of Proceedings of the National Academy of Science (still to be obtained)

Table 1 Average uncorrected pairwise distances between and within populations of Barbary macaques in North Africa and inferred divergence times (t) (after Modolo *et al.*, 2005).

Comparison	*% divergence*	*SD*	$t_{divergence}$
Algeria:Morocco	4.43	0.62	1.24-1.85 mya
Kherrata:Morocco	4.10	0.32	1.22-1.62 mya
Akfadou:Djurdjura	2.60	0.18	0.780-1.01 mya
Kherrata	0.59	0.40	64-363 kya
Morocco	0.55	0.22	110-282 kya
Akfadou	0.21		70-78 kya
Djurdjura	0.21		70-78 kya

Note that divergence time $t_{divergence}$ was based on an initial speciation time for *Macaca* of 5.5 mya (Delson, 1980; Tosi *et al.*, 2003). Although the absolute timing of divergence is questionable because direct calibration on the fossil record can lead to different results, at least the sequence and timing of event convincingly reflect the evolutionary history.
mya, million years ago; kya, thousand years ago

haplotype presumably expanded from the Middle Atlas northward and southward into the Rif and High Atlas mountains, respectively. Moreover, haplotype number M08, found exclusively in the Rif, is connected to a haplotype which is derived from the central haplotype. This suggests that there must have been at least a second episode of emigration from the Middle Atlas into the Rif mountains, possibly indicating that forest corridors were present in Morocco for a longer period than in Algeria. These migrations were presumably possible due to interconnected extended forests until the beginning of forest exploitation through human activities around 2200 years ago (Lamb *et al.*, 1989).

Generally speaking, Algerian and Moroccan lineages are very distinct in genetic terms, although habitats in Morocco and Algeria might have been interconnected through extensive forests until the beginning of forest exploitation in Roman times. Surprisingly, Moroccan populations do show mtDNA variation, whereas variation within Algerian populations is virtually non-existent. A low level of variation may be explained by female philopatry on one hand; but on the other hand, these patterns may also arise from bottlenecks within each population. Djurdjura is estimated to be the largest remaining Algerian population, yet only two haplotypes were found. Sampling took place in five different social groups, thus excluding the possibility that sampling was conducted within the geographical boundaries of a single genetic stock. This indicates that the pattern of haplotype variation within each population has been shaped not only by climatic changes but also by comparatively recent bottlenecks, caused by deforestation or other human disturbances.

To date, the genetic structure of natural populations of *Macaca sylvanus* has been the focus of only a few studies. Scheffrahn *et al.* (1993) conducted a genetic survey in two Algerian populations (Akfadou and Djurdjura), based on electrophoretic analysis of five blood protein loci. The Djurdjura group experienced group fission into three stable groups during the study period, but no significant genetic subdivision among the new groups was detected. In a more recent study of the same population however, microsatellite analysis revealed genetic distance F_{ST} values indicating significant genetic subdivision among the social groups in Djurdjura (Modolo, 2005). Moreover, von Segesser *et al.* (1999) not only found significant levels of differentiation in this particular population, but also determined that all four Algerian isolates were genetically distinct. Most probably, barriers to migration such as numerous intervening built-up areas with no forest patches have prevented further gene flow. Dispersal seems to be limited even between the social groups within the isolate of Djurdjura. Surprisingly, there was no evidence of reduced allelic diversity within the Algerian populations despite the small population size for some of the isolates (von Segesser *et al.*, 1999). Based on the results of microsatellite analysis, it would seem that forest fragmentation in Algeria has not yet led to a detectable reduction in genetic diversity, although there is significant genetic subdivision between the remaining isolates. Furthermore, there was no demonstrable difference in allelic diversity between small and large isolates. Obviously, the currently isolated subpopulations have undergone some degree of independent genetic differentiation and population size was large enough to retain this genetic variation at least in the short term (von Segesser *et al.*, 1999). However, results from mtDNA analysis reveal a different conclusion in respect to genetic diversity (Modolo *et al.*, 2005). Algerian isolates are, in fact, genetically highly distinct, but variation within the subpopulations is virtually non-existent. This discrepancy is partially attributable to differences in inheritance of genetic markers and dispersal patterns (Melnick and Hoelzer, 1992). Maternal inheritance of mtDNA combined with female sedentism in Barbary macaques can lead to local fixation of divergent lineages, even in the absence of geographical barriers (Melnick *et al.*, 1993). However, observed levels of within-population mtDNA diversity in Algeria are reduced to such an extent that they possibly derive from recent bottleneck events rather than from the mode of inheritance of the marker. The Arab conquest and the immigration of nomads intensified forest destruction, with the result that by the end of the 17th century most of the lowlands had been cleared. This could have caused bottlenecks in population size of Algerian populations and hence to reduction in genetic variation to the very low level that is now observed.

Barbary macaques in Gibraltar

The origin of the Gibraltar Colony

Following an inferred origin in northern Africa at least 5.5 mya, early macaques presumably entered southern Europe by the beginning of the Pliocene (Delson, 1980; Martin, 1990). *Macaca sylvanus* apparently flourished in Europe only during the warm phases of the Pliocene and Pleistocene, persisting at least until the last interglacial and then eventually retreating from the continent entirely (Delson, 1980). The question of the origin of Gibraltar's Barbary macaques has generated considerable speculation. Although Gibraltar has a very rich Pleistocene fauna, no fossil macaques have ever been discovered there. One hypothesis suggested that Moorish people imported Barbary macaques from North Africa during their occupation of southern Spain between 711 and 1462 AD, but definite written records of their presence first appear in 1704, when an Anglo-Dutch fleet captured the Rock (Morris and Morris, 1967). Whatever its origin, however, the Gibraltar population has been the subject of intense human influence ever since it was founded. For a long period of time, the British army was in charge of provisioning the macaques and controlling population size by removing surplus animals. When the number of animals was too low, the colony was restocked several times with individuals from North Africa. After the last population bottleneck at the time of the Second World War, the most recent introductions in the period 1938-45 took place from Morocco (Fa, 1984a).

Repeated introduction of animals and the lack of reliable data concerning founders of the present-day population have obscured the origin of the Gibraltar colony. In order to reconstruct the geographical origin of the present population, the most variable segment of the control region of mtDNA of one third of the Gibraltar macaques was sequenced and compared to haplotypes of natural populations (Modolo *et al.*, 2005). Analysis of the origin of the Gibraltar colony was possible because Algerian and Moroccan mtDNA haplotypes were found to be distinct and they permitted allocation of haplotypes to specific North African subpopulations. The investigation revealed that all extant Gibraltarian haplotypes were also found in North Africa, thus greatly diminishing the possibility that the Gibraltar macaques represent or include any remnant of the original European population, although the possibility that certain previous Gibraltarian lineages may have already become extinct cannot be excluded. Figure 5 illustrates the affiliation of haplotypes found in Gibraltar with current Moroccan and Algerian haplotypes. Congruent results were also found using microsatellites (von Segesser, 1999; Modolo, 2005). The Gibraltar population as a whole now possesses only three mtDNA haplotypes. Probable sources of the present Gibraltar colony are the nearby Rif region in northern Morocco and the Akfadou population in Algeria. The fact that each haplotype could be assigned to a particular wild population may be important when considering reintroduction of animals into the wild (see section below).

Genetic variability

Population size reduction and fragmentation are expected to have major genetic consequences for a population. Nowadays, however, population fragmentation and isolation are usually the consequence of long-lasting declines, triggered by habitat loss or other factors related to human activities (Caughley, 1994). As already described above, the Gibraltar macaque colony was subject to intense human intervention and population size was mainly regulated by removing individual animals. The fact that until 1980 population size was actively limited to about 35 animals (Fa and Lind, 1996) had a strong impact on this colony. Furthermore, the way this management strategy was implemented, namely by removing mainly males and leaving only one adult male in each group, reduced the effective population size still further.

Contrary to expectations, earlier work on the genetic variability of the Gibraltar colony revealed no detectable reduction in genetic diversity compared to wild-living Barbary macaques (von Segesser, 1999). There was, however, clear indication of homozygous excess, which can result from inbreeding, assortative mating or a Wahlund effect, when alleles become fixed in subpopulations. Additionally, F_{ST} estimates revealed no significant genetic subdivision between social groups, indicating high levels of gene flow between them. However, a recent study with a much larger sample size has yielded different results (Modolo, 2005). It seems that long-term isolation and multiple bottlenecks have, in fact, led to a significantly lower allelic variability and expected heterozygosity in Gibraltar macaques compared to unmanaged wild populations in Algeria and Morocco. Moreover, each social group in Gibraltar is significantly differentiated in genetic terms from the other groups, and levels of population differentiation are comparable to genetic structuring found in natural Algerian social groups. Despite the much larger sample size in this study, private alleles in the Gibraltar population were found to be present at lower frequency than in Algerian and Moroccan populations. This is somewhat surprising; considering the limited habitat area in Gibraltar, one would expect an increased level of gene flow between social groups. However, it seems that non-random mating and reduced male migration rates (Modolo, 2005) compared to wild populations (Mehlmann 1986; Ménard and Vallet 1993) might be responsible for these results. The discrepancy between these two studies on Gibraltar is probably attributable to the fact that von Segesser (1999) had a very limited sample size (n = 24) and used only six microsatellite loci compared to the subsequent study, in which 127 individuals and an increased set of genetic markers (14 microsatellite loci) with extensive statistical analysis were applied. However, both studies clearly show that human intervention has had a major impact on the genetic variability of the species. It was concluded by von Segesser (1999) that there was still time to preserve allelic diversity in this isolated population, but she emphasised the need to develop viable long-term management strategies for the species to slow the rate of genetic drift.

Captive populations

Breeding of endangered species in captivity has become an important issue for declining primate populations in the wild. One major goal is eventual reintroduction into their natural habitats or into an acceptable similar environment, in order to re-establish free-living populations to a previous range. On the other hand, such captive populations also represent a valuable source for behavioural studies or education. Clearly, captive environments differ significantly from wild conditions, and the restricted gene pool represented by the (often) small number of individuals kept in captivity is particularly susceptible to deleterious changes. Maintaining genetic diversity in captive populations has often been emphasised as being the most critical element for the success of captive breeding programmes.

The largest captive colonies of Barbary macaques are at Kintzheim and Rocamadour in France and that in the outdoor enclosure at Salem, Germany (Figure 3). All three colonies were founded in the seventies with Barbary macaques taken from the Azrou region in the Moroccan Middle Atlas. Animals are kept outdoors throughout the year and are artificially provisioned. Because of the abundant food provisioning, population size may grow faster than in the wild. For this reason, females receive contraceptive implants to limit breeding.

Although numerous studies have been carried out in Salem (e.g. Paul and Küster, 1988; Küster *et al.*, 1994; Paul and Küster 1996), including paternity analysis (Küster *et al.*, 1992), they were mainly directed toward behavioural questions and none of them addressed genetic diversity. The genetic study by von Segesser (1999) included a group of the Kintzheim colony (51 samples), three social units of the Salem colony (20 samples) and 10 individuals from the Beauval Zoo in France. The author used five microsatellite loci and compared results with those obtained from wild populations of Algeria and Morocco. It emerged that no semi-captive population analysed showed a significant reduction in heterozygosity or in allelic diversity. This finding was explained by the relatively brief period over which these macaques have been separated from their original location combined with the long lifespan of the species. However, none of these colonies was found to be in Hardy-Weinberg equilibrium, probably as a consequence of human interventions, such as removal of specific animals, and as a result of outbreeding depression due to reproduction of individuals from widely separated subpopulations.

Implications for Conservation Management

Human activities are having a devastating effect on the survival of natural populations. Loss of habitat, reduction in population size and changes in the connectivity of populations due to human disturbances are currently the most important factors causing loss of biodiversity (Frankham, 1999). Genetic management of threatened species is still in its infancy, but consideration of genetic aspects when developing action plans is essential. The needs of the local human population should also be included in any plans in order to guarantee an effective and continuous implementation.

Wild populations

Over many centuries, unmanaged exploitation of the forests in Morocco has resulted in considerable loss of biodiversity and land degradation (Mikesell, 1960). The primary cause of environmental degradation is the rapid growth of the human population, which creates pressure on forest use and on the agricultural system in general. Additionally, a sustained increase in livestock in that region has increased the competition for resources. Barbary macaques are directly affected by these factors because of their dependence on specific habitats that are undergoing degradation and fragmentation. Productive alternatives should be proposed to local farmers and herdsmen, in order to slow down the ongoing damage, although provision of alternative market opportunities and establishment of a minimum income are difficult objectives to achieve. At the same time, the establishment of protected and fenced areas in the remaining forest patches are urgently needed to restrict the access of grazing animals (Camperio Ciani *et al.*, 2005). These protected areas would secure habitat for the Barbary macaques and protect the forests from further degradation. The local inhabitants should be provided with information about sustainable use of the forest, and their needs should be explicitly incorporated into conservation plans. Educational efforts should create respect for the Barbary macaques in order to reduce their exploitation and further their protection.

Even when protected areas are successfully established, there is a need for additional action to maintain population viability. Because of isolation and intervening built-up areas between Barbary macaque populations in forest fragments, genetic exchange through male dispersal is not possible. It has been shown that habitat corridors may be valuable conservation tools (Beier and Noss 1998). While some species readily move between fragments using habitat corridors, others do not. It is not clear whether Barbary macaques would use corridors, nor how such corridors should be constituted. Further investigations are needed to determine how habitat connections are used, if at all. Given the close relationship among Moroccan mitochondrial haplotypes, it is conceivable that the small populations in the Rif and High Atlas could be restocked with individuals from the Middle Atlas, although the establishment of protected areas would be an essential precondition.

In Algeria, pressure on forest use and fragmentation of habitats is operating as in Morocco. The remaining populations are completely isolated by distances of 50-200 km. Natural genetic exchange between the isolates is ruled out because of numerous built-up areas and discontinuity of forest patches. Although numerous studies on Barbary macaques in Algerian populations have been conducted in the past, the present situation remains uncertain because all field research was halted more than a decade ago as a result of political instability. As in Morocco, grazing animals such a sheep and goats are direct competitors for Barbary macaques in habitat use. Forest degradation as a consequence of underbrush reduction caused by livestock is presumably affecting the macaque population as in Morocco. Given the fact that all extant Algerian macaque populations are genetically distinct, future guidelines should aim at overall protection of existing populations. Algerian individuals should not be translocated between subpopulations, because of the relatively large genetic distances separating them, but only within any given population. Outbreeding depression should be considered as a risk to conservation of such deeply subdivided populations. Outbreeding depression is defined as a reduction in fitness of hybrid populations caused by the disruption of locally adapted gene complexes (Frankham, 1999). Currently, conservation genetics advocate an extremely cautious approach towards mixing populations, although there is apparently little evidence that outbreeding depression is important in mobile animals (Frankham, 1995b). It may therefore be questioned whether the current approach of not mixing populations is justifiable in the face of fragmented populations on their way to extinction. However, the wisest course is for conservation efforts for Algerian populations to be reinforced in order to preserve their unique genetic lineages as far as possible, particularly for the smaller populations. Conservation interests may be achieved most efficiently not by translocation of animals between populations, but rather by halting the reduction of population size and by creating protected areas and corridors within a given population.

In view of the ever-increasing rate of habitat fragmentation and loss, there is an urgent need for enhanced conservation efforts in both Morocco and Algeria. We are only just beginning to understand the complex relationships between primates and the damaged and fragmented forests they inhabit.

Captive populations

Analyses conducted in the course of a recent study have shown that, for all measures of genetic diversity, the Gibraltar colony showed uniformly low values at all sites compared to unmanaged wild populations in Morocco and Algeria (Modolo, 2005). Overpopulation has always been the prime reason for regulating population size in the Gibraltar colony. Reduction was achieved by removing aggressive and dispersing males rather than randomly chosen individuals, hence inadvertently applying strong and relentless selective pressure against aggressiveness and dispersal. Such artificial and selective intervention should be avoided in the future, particularly because it increases the risk of genetic drift and does not reduce the number of offspring. In response to overpopulation, management actions should consider the selective and long-term application of contraceptive implants, as currently conducted in several outdoor enclosures (Salem, Kintzheim, Rocamadour; Paul and Küster 1988; de Turckheim and Merz 1984). Each female should be allowed to breed at least once in her lifetime, in order to improve her social integration into the group and to maintain genetic diversity. Because of the long lifespan of the species, such actions would not serve to reduce population size in the short term, but they would at least minimise the hitherto persistent problem of overpopulation. As another alternative, introduction of surplus animals into wild habitats in North Africa, from which they have disappeared, could be taken into consideration, in view of the threatened status of the species. However, it must be borne in mind that the Gibraltar macaques have a mixed ancestry, and this renders translocation into appropriate populations problematic. Possible outbreeding depression caused by breeding between distantly related individuals is best avoided in conservation actions. Nevertheless, because social groups in Gibraltar are subdivided according to haplotypes and significant population structuring was also found for autosomal markers, it is conceivable that Gibraltar macaques are effectively separated according to their origin. Thus, it might be less problematic to consider reintroduction into wild populations, provided that the inferred population of origin is taken into account.

Last but not least, in order to reduce effects of isolation further, translocation of single individuals from other outdoor enclosures into Gibraltar should be considered. In a study of an insular population of song sparrows, it was shown that a few immigrants that entered the population each year were sufficient to maintain genetic diversity (Keller *et al.*, 2001). Dispersal appears to be an important factor in maintaining genetically viable populations, although simple models regarding the correct number of migrants may not apply to all populations. Male Barbary macaques do normally disperse when they reach maturity, usually into neighbouring groups. Management actions would include removal of subadult or young adult males from other outdoor enclosures, preferentially taking individuals that have already made attempts to migrate into other groups, and release of these males on Gibraltar after the mating season when aggression among males is lower. It may be possible to transfer specific individuals manually between Gibraltar groups. Because novel males in social groups are rare, females should have a strong preference for them.

Acknowledgements

I am very grateful to everybody who kindly helped in sample collection or provided samples: F. Botte-von Segesser, K. Hodges, R. Kümmerli, N.

Ménard, U. Möckli, U. Möhle, C. Roos, W. Scheffrahn. I thank John Cortes, Mark Pizarro and Eric Shaw for research permission and assistance in sampling in Gibraltar, M. Bruford, J. Pastorini and H. Zischler for laboratory help and R. Martin for his guidance throughout this work and helpful comments on the manuscript. The research has been supported by grants from the Swiss National Foundation and the A. H. Schultz Foundation (Zürich).

References

Amos W and Balmford A (2001) When does conservation genetics matter? *Heredity* **87** 257-265

Beebee T and Rowe G (2004) *An introduction to molecular ecology* Oxford University Press, New York

Beier P and Noss RF (1998) Do habitat corridors provide connectivity? *Biological Conservation* **12** 1241-1252

Caccone A, Gentile G, Gibbs JP, Fritts TH, Snell HL, Betts J and Powell J (2002) Phylogeography and history of the giant Galapagos tortoises *Evolution* **56** 2052-2066

Camperio Ciani A (2003) La desertificazione in Marocco: uso degli indicatori biologici nel monitoraggio della desertificazione delle foreste del Medio Atlante *Antropologia Mediterranea* **1** 57-68

Camperio Ciani A, Palentini L, Arahou M, Martinoli L, Capiluppi C and Mouna M (2005) Population decline of *Macaca sylvanus* in the Middle Atlas of Morocco *Biological Conservation* **121** 635-641

Caughley G (1994) Directions in conservation biology *Journal of Animal Ecology* **63** 215-244

de Turckheim G and Merz E (1984) Breeding Barbary macaques in outdoor open enclosures. In *The Barbary macaque: a case study in conservation* pp 241-261 Ed JE Fa. Plenum Press, New York

Delson E (1980) Fossil macaques, phyletic relationships and a scenario of deployment. In *The Macaques: Studies in ecology, behavior and evolution* pp 10-30 Ed DG Lindburg. Van Nostrand Reinhold, New York

Fa JE (1984a) Habitat distribution and habitat preference in Barbary macaques (*Macaca sylvanus*) *International Journal of Primatology* **5** 273-286

Fa JE (1984b) Structure and dynamics of the Barbary macaque population in Gibraltar. In *The Barbary macaque: A case study in conservation* pp 263-306 Ed JE Fa. Plenum Press, New York.

Fa JE and Lind R (1996) Population management and viability of the Gibraltar Barbary macaques. In *Evolution and ecology of macaque societies* pp 235-261 Eds FE Fa and DG Lindburg. Cambridge University Press, Cambridge

Frankham R (1995a) Inbreeding and extinction: a threshold effect *Conservation Biology* **15** 792-799

Frankham R (1995b) Conservation genetics *Annual Review of Genetics* **29** 305-327

Frankham R (1997) Do island populations have less genetic variation than mainland populations? *Heredity* **78** 311-327

Frankham R (1999) Quantitative genetics in conservation biology *Genetical Research* **74** 237-244

Hayasaka K, Fujii K and Horai S (1996) Molecular phylogeny of macaques: Implications of nucleotide sequences from an 896-base pair region of mitochondrial DNA *Molecular Biology and Evolution* **13** 1044-1053

Küster J, Paul A and Arnemann J (1992) Paternity determination by oligonucleotide DNA fingerprinting in Barbary macaques (*Macaca sylvanus*). In *Paternity in primates: Genetic tests and theories* pp 141-154 Eds RD Martin, AF Dixson and EJ Wickings. Karger, Basel

Küster J, Paul A and Arnemann J (1994) Kinship, familiarity and mating avoidance in Barbary macaques, *Macaca sylvanus Animal Behaviour* **48** 1183-1194

Lamb HF, Eicher U and Switsur VR (1989) An 18,000-year record of vegetation, lake-level and climatic change from Tigalmamine, Middle Atlas, Morocco *Journal of Biogeography* **16** 65-74

Martin RD (1990) *Primate origins and evolution* Chapman Hall/Princeton University Press, London

Mehlman PT (1986) Male intergroup mobility in a wild population of the Barbary macaque (*Macaca sylvanus*), Ghomaran Rif Mountains, Morocco *American Journal of Primatology* **10** 67-81

Melnick DJ and Hoelzer GA (1992) Differences in male and female macaque dispersal lead to contrasting distributions of nuclear and mitochondrial DNA *International Journal of Primatology* **13** 379-393

Melnick DJ, Hoelzer GA, Absher R and Ashley MV (1993) mtDNA diversity on rhesus monkeys revelas overestimates of divergence time and paraphyly with neighbouring species *Molecular Biology and Evolution* **10** 282-295

Ménard N and Vallet D (1993) Dynamics of fission in a wild Barbary macaque group (*Macaca sylvanus*) *International Journal of Primatology* **14** 479-500

Mikesell MW (1960) Deforestation in Northern Morocco *Science* **132** 441-448

Modolo L (2005) *Phylogeography and population genetics of Barbary macaques* (Macaca sylvanus) Ph.D. thesis, University of Zürich

Modolo L, Salzburger W and Martin RD (2005) Phylogeography of Barbary macaques (*Macaca sylvanus*) and the origin of the Gibraltar colony *Proceedings of the National Academy of Sciences USA* **102** 7392-7397

Morris R and Morris D (1967) *Men and apes* Hutchinson and Co., London

Paul A and Küster J (1988) Life-history patterns of Barbary macaques (*Macaca sylvanus*) at Affenberg Salem. In *Ecology and behavior of food-enhanced primate groups* Eds JE Fa and CH Southwick. Alan R. Liss, New York

Paul A and Küster J (1996) Infant handling by female Barbary macaques (*Macaca sylvanus*) at Affenberg Salem: testing functional and evolutionary hypotheses *Behavioral Ecology and Sociobiology* **39** 133-145

Saccheri I, Kuussaari M, Kankare M, Vikman P, Fortelius W and Hanski I (1998) Inbreeding and extinction in a butterfly metapopulation *Nature* **392** 491-494

Scheffrahn W, Ménard N, Vallet D and Gaçi B (1993) Ecology, demography, and population genetics of Barbary macaques in Algeria *Primates* **34** 381-394

Simmons MJ and Crow JF (1977) Mutations affecting fitness in *Drosophila* populations *Annual Review of Genetics* **11** 49-78

Spielman D, Brook BW and Frankham R (2004) Most species are not driven to extinction before genetic factors impact them *Proceedings of the National Academy of Sciences USA* **101** 15261-15264

Taub DM (1977) Geographic distribution and habitat diversity of the Barbary macaque *Macaca sylvanus* L. *Folia Primatologica* **27** 108-133

Taub DM (1982) A brief historical account of the recent decline in geographic distribution of the Barbary macaques in North Africa. In *The Barbary macaque: A case study in conservation* pp 71-78 Ed JE Fa. Plenum Press, New York

Tosi AJ, Morales JC and Melnick DJ (2003) Paternal, maternal, and biparental molecular markers provide unique windows onto the evolutionary history of macaque monkeys *Evolution* **57** 1419-1435

von Segesser F (1999) *Conservation genetics of Barbary macaques* (Macaca sylvanus) Ph.D. thesis, University of Zürich

von Segesser F, Ménard N, Gaçi B and Martin RD (1999) Genetic differentiation within and between isolated Algerian subpopulations of Barbary macaques (*Macaca sylvanus*): evidence from microsatellites *Molecular Ecology* **8** 433-442

Distribution and demography of the Barbary Macaque (*Macaca sylvanus* L.) in the wild

M Mouna[1] and A Camperio Ciani[2]
[1]*Département de Zoologie et d'Ecologie Animal, Institut Scientifique, B. P. 703, Rabat 10106, Morocco*
[2]*Dipartimento di Psicologia Generale, Universita' Degli Studi di Padova, Via Belzoni, 80 - 35131 Padova, Italy*

Introduction

The earliest fossil records of *Macaca sylvanus* dating back 7MY to the Miocene can be found in Egypt. Subsequently they occur almost everywhere in the Mediterranean basin and in Europe, corresponding to the Plio-pleistocene. The maximum distribution, based on fossil findings of forms related to the modern *Macaca sylvanus* was during the late Pleistocene and ranged from southern England in the north, to Turkey and the Mediterranean coast from Syria to Morocco (Delson, 1980; Camperio Ciani , 1986).

Following the late pleistocenic expansion the distribution of *Macaca sylvanus* progressively declined. Currently, *M. sylvanus* still can be found in the cedar and oak forests north of the Sahara, in Morocco and Algeria, and these now constitute the last remaining populations of the species in the wild. Based on the most recent studies, we now estimate a total population size of fewer than 10,000 individuals. This dramatic decline in population size from the estimated 20-25000 reported by Fa *et al.* (1984) has stimulated research into the ecology and demography of the species in an attempt to provide a database useful for developing plans for *in situ* management of remaining populations. Our own study was carried out in the Middle Atlas region of Morocco between 1994 and 2002. Here we present the results of this field study and compare them with other data, both published and unpublished. We will focus on the distribution of the species in the wild, its habitat, demography and the principal causes of the decline of its populations both in Morocco and in Algeria. We will conclude with some recommendations on how best to preserve these wild animals in what is left of their natural ecosystem.

Former and current distribution

Several fossil records of *Macaca sylvanus*, during the time of the largest expansion of the species at the end of Pleistocene (100,000-50,000 ybp), have been found distributed through much of Europe and in North Africa from Morocco to Egypt (Camperio Ciani, 1986). Since then the macaques disappeared from the northern coasts of the Mediterranean, then from Sardinia, and, in the last few centuries, also from Turkey. The end of the 19th century saw the disappearance of the Barbary macaque east of Tunisia and by the beginning of the 20th century, *M. sylvanus* had completely disappeared except from the scattered refuges in Morocco and Algeria (Taub, 1984, Camperio Ciani, 1986). Nowadays macaques persist mainly in the remaining forests of pure cedar or cedar mixed with oaks, although some smaller populations are now taking refuge in steep rocky areas.

In Morocco, *M. sylvanus* is found only in the northern mountain chain of the Rif, in the Middle Atlas and in the High Atlas mountains in the south. The distribution ranges in altitude from sea level in Mediterranean coastal forests up to 2200m in the central Middle Atlas temperate cedar forests.

In the Rif, the very fragmented populations of the Barbary macaque are distributed in several localities: Jbel Moussa, Fahs Lemhar (Belouazen), Jbel Bouzaïtoun, Jbel Kelti, Kaiat, Jbel Sidi Salah, Jbel Bouhachem, Jbel Tazoute, Beni Mhamed, Jbel Tissouka, Jbel Laqra, Jbel Talassemtane, Oued Tijidda, Oued Adelmane and Jbel Tizirene. Altogether, this represents a total population of 400-800 individuals (Fa *et al.*, 1984). The individual populations are distributed along an axis running from the North to the East (Figure 1) following mountainous peaks or in steep rocky areas. In the North, four groups of approximately 90 Barbary macaques still exist at Jbel Moussa (Sehar, 2004)*, although only 11 individuals were reported in 1980 (Fa *et al.* 1984) and 8 individuals observed in a more recent study in 2005 (Waters *et al.*, in press). A small colony of animals located South of Tétouan, still exists in the rocky escarpments of Jbel Ghorghez (5-10 individuals) **. Finally, a recently-discovered population of seven groups (approximately 400 individuals) exists in the East of the Rif at Jbel Gourougou (in the region of Nador) ***.

In the Middle Atlas, the Barbary macaque can be found in the regions of Taza (Tamjilt, Taffert, Tahafourt), Ifrane - Azrou - Michlifène, Aïn Leuh - El Hammam, of Seheb, of Ajdir and Midelt (Fa *et al.*, 1984). Following recent searches, we can add reports of the presence of macaques on the northern

---

*Found on escarpments and littoral cliffs in Jbel Moussa - Jbel Snissel, upstream of Lghar Lakhal, cliffs of Toura - Ras Léona and Jbel Jouima.

**Observed in 2001, in a dump close to Aïn Bouânane and are often encountered in the mountains by local inhabitants (personal communication, 2003).

***Distributed on Oulad Assdi Issa (neighbourhood of the top of Jbel Tigria), northern slopes of Jbel Tigria (close to the source), the forest of Karmoud, close to Ikhannoussane, Tifarouine - Ch' hiba Jbel Bayyou, col of Aghilass and Tazouda (Sehhar, 2003).

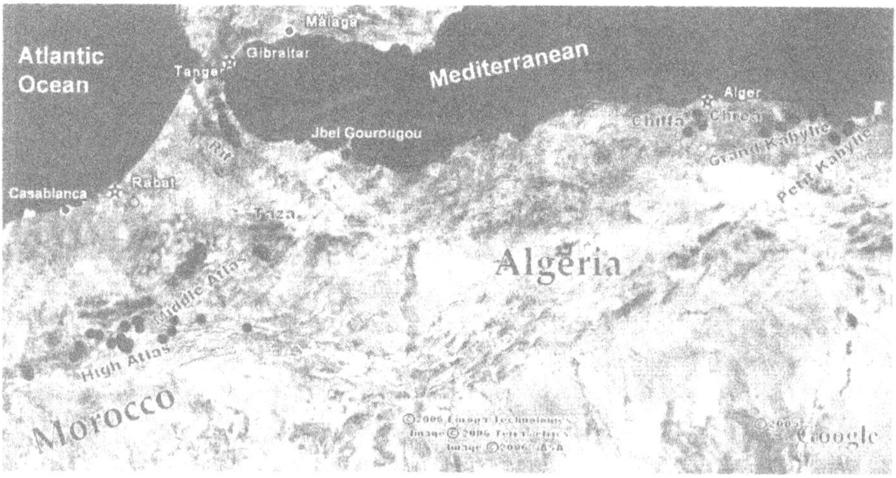

Figure 1. Map of northern Morocco and Algeria (Google Earth™ mapping service) and the distribution of *Macaca sylvanus* in the wild (●).

slope of the mountain mass of Beni Mellal, south of Zaouat Echikh, the upper and the middle basin of Oued El Abid and the cascades of Ouzoud (Cuzin, 2003). This represents four previously unrecorded small populations totalling approximately 100-200 individuals.

In the High Atlas, Fa *et al.* (1984) only mentioned the existence of macaques in the Ourika valley, whereas they have been widely distributed throughout the region during the last twenty years. In fact, monkeys were observed personally (ACC) in the basins of the Oueds Ourika (100-200) and upper Zat region of Marrakech, and in the basins of Oueds Ghassat, Tifni, Lakhdar and Ouhansal (Tamga: High Atlas)* (by MM and ACC) which we estimate to represent 200-300 individuals, including 100-120 in Tamga. In addition, macaques have been reported in Assif Melloul (downstream of Anergui), in cedar forests in the west of Tounfit and in Tagouilelt (foot of Jbel Ayachi) by Cuzin (2003).

In Algeria, the Barbary macaques could be found prior to 1984 in 7 localities: Chiffa, Grande Kabylie (Bejaia, Djurdjura and Aqfadou) and Petite Kabylie (Kerrata, Babors and Guerrouch) (Fa *et al.*, 1984). Currently, the species is scattered in other places, particularly in the national park of Chrèa (Blidèen Atlas), where 4 groups can still be found, although their origin and number are unknown (Dehale et Chakali, personal communication, 2005)**.

---

*At Tamga, the species occurs in the escarpments and gorges of Oued Ouhansal and Assif Melloul in particular in the passes of Akhachane, the cliffs of Ouakhouden, Tissili Khlif (gorges of Ouhansal between Tazoult and Jbel Waouriroute) and Tadawt Tiritine (gorges of Ouhansal idownstream of Tamga) (Sehhar, 2004).

**Located at Chiffa, El-Hamdania, Beni-Ali (new group) at an altitude of 1200m and in Magtaa El-Azrag (a recently discovered group) (Dehale et Chakali, pers.comm., 2005).

Habitat of Barbary macaques

Barbary macaques live both in Morocco and Algeria, in cedar forests, oak forests and in particular in the mixed cedar - oak forests which appear more varied in undergrowth, biodiversity and consequently richer in food. Nevertheless, macaques can still find refuge in a multitude of areas like Mediterranean patch forests, clear or dense forests composed of pure or mixed pine, fir forests (Rif, Morocco), cedar, oak, mixed forest (Middle Atlas: Morocco) and other habitats such as chestnut-groves (Chrèa: Algeria). *Macaca sylvanus* is predominantly a ground forager and feeds on a large variety of roots, buds, shoots, fruits and seeds and searches under stones for invertebrates such as scorpions, ants and insect larvae (Drucker, 1984). During the winter, when snow covers the ground, they feed on cedar shoots, and in the autumn on acorns. In the summer, in very dry conditions, if they cannot access water sources (most populations are surrounded by permanent human settlement), they occasionally bark strip cedar tree branches and tops to get liquid and nutrients, but this happens only in the Middle Atlas of Morocco (Camperio Ciani *et al.*, 2001). Overall, Barbary macaques prefer habitats with tall trees, not just for feeding during periods of snow, but for security. They mostly sleep overnight on high trees in small family clusters. Adults spend a large part of the daytime on the ground but rapidly climb into tall trees when alarmed. The most common predators are free ranging dogs, jackals and foxes, but even humans hunt them for sale to the tourist market. The best defensive strategy these animals have against such large predators is to climb on top of trees or take refuge on steep cliffs.

Although macaques are also found in scrub vegetation (High Atlas: Morocco), they always occur close to a refuge such as tall trees or cliffs (Cuzin, 2003). A small proportion of the Mediterranean coastal population at present lives in steep rocky areas including littoral cliffs where the herbaceous vegetation is inaccessible to domestic animals, although this is probably a secondary adaptation to the loss of forested habitat in these areas. In both Morocco and Algeria, these probably represent the most threatened populations due to their very limited size and the ever growing human presence nearby (Ménard *et al.*, 1985; Ménard *et al.*, 1990; Cuzin, 2003). Further details on specific habitats are given in Fa 1984, and Camperio Ciani and Mouna, this volume.

Demography of the Barbary Macaque

Morocco

According to the most recent estimates (Camperio Ciani *et al.*, 2005), between 8000 and 10000 monkeys currently live in Morocco, including approximately 6000 animals in the Middle Atlas, which constitute the most important

Cedars near the lake of Aguelmane Afennourir, Middle Atlas, after branches have been removed to provide fodder for sheep and goats during the winter. Copyright J. Cortes

population in the species' entire range. According to Mehlman (1984) the populations of the Rif 20 years ago were of a younger structure with apparently less post-natal loss (the colony in Ghomara) than in the Middle Atlas. Moreover Lavieren (2005) recently reported a decrease in the numbers of infants of 29.2% in August/ September and up to 68% in November/ December from the Middle Atlas regions of Ifrane and Azrou, a degree of loss which could be highly significant in relation to the stability of the population in this region.

In addition, an average of only approximately 24% immature animals (i.e. less than 5 years old) was reported for populations at Aïn Kahla in the

Middle Atlas (Ménard, 2002) against 48 - 56% at the same site twenty years ago (and in cedar forests in general). Birth rates (*i.e.* number of young per female) at this site apparently varied from 56 to 100% and the survival of the infants beyond their first year was between 80 and 87% (Deag, 1974; Taub, 1977, 1978 in Ménard, Vallet et Gautier-Hion, 1985). According to Mehlman (1984), the figure of 24% of immature animals observed in the Middle Atlas, is even lower than that found in steep rocky areas and can be expected to eventually result in a reduction in population size and density. Indeed, whereas Deag (1974) noted a density of 44 animals per square km in the late 1960s, this figure had already fallen to only 24.9 animals per square km in 1994 - 1995 (Mouna *et al.*, 1999).

In 2002, macaque density in the Middle Atlas was found to be only 10 animals per square km (Camperio Ciani *et al.*, 2004). Furthermore, compared to the earlier studies, the Barbary macaque in the Middle Atlas now seems to present a profile of an unbalanced population with an abnormally high number of adult males, whereas the number of females has not shown much change (Deag, 1974; Camperio Ciani *et al.*, 1999; Mouna *et al.*, 1999) (Table 1).

Table 1. Comparison of the structure of the populations of the Barbary macaque in the Middle Atlas

Study	*Males*	*Females*	*Immature*	*Density (animals/km²)*
Deag, 1974	26,20	24,50	49,30	44
Current study, 1995 and 1998*	39,04	24,60	36,35	19,56

* Average of the two years

In the Rif, the population of the Barbary macaque currently living in Jbel Moussa comprises approximately 90 individuals, represented by four groups. (Sehhar, 2004). In Jbel Gourougou, also in the Rif, the introduction of 82 monkeys (75 females and 7 males) in 1985 has resulted in a population of approximately 400 animals made up of seven groups of 40 to 80 animals. Here, monkeys have been introduced into an area close to human settlement and it is no surprise that they cause damage to leguminous crops and fruit trees and are subject to persecution by the local people and by roaming dogs (Sehhar, 2003).

In the High Atlas, the populations of Barbary macaques are so fragmented that movement between colonies has become difficult or impossible, with the result that gene flow between them is extremely limited (Scheffrahn *et al.*, 1993).

Considering the current state of macaque habitats and their transformation by Man, Cuzin (2003) distinguished three populations of macaques in the southern (High) Atlas region:

- populations at high risk in Tagouilelt (probably already extinct), in the middle Oued El Abid and in the northern Oued Ourika (which are most probably of small numbers not exceeding one hundred individuals).
- populations at moderate risk in lower Oued El Abid - Ouzoud, Oueds Ghassat and Tifni basins, upper Oued Lakhdar, southern Oueds Ourika - Zat;
- populations at reduced risk in upper Oued El Abid, middle Oued Ouhansal - Assif Melloul, lower Ouhansal - gorges of Wabzaza, National Park, eastern High Atlas - Aqqa n'Ouanine.

In general, groups of macaques comprising 2 to 50 animals were observed both in the High Atlas and the southern Middle Atlas (Cuzin, 2003). In Tamga (mainly Oued Ouhansal and Assif Melloul in the region of Azilal in the High Atlas), the current population of the Barbary macaque consists of four groups of 20 to 40 individuals, totaling approximately 120 animals, but highly threatened by overgrazing and forestry exploitation.

Algeria

The last estimate of the total number of macaques living in Algeria dates to the early 1980's and varies between 5000 and 6200 individuals (Fa *et al.*, 1984). Information on the demography of the Barbary macaque in Algeria was provided by Ménard *et al.* (1985) from studies carried out in forest sites and in steep rocky areas. The authors reported that climatic changes and the negative impact of Man on the habitat have acted unfavorably on the populations of the monkey during the last fifty years, with fragmentation being one of the main effects. Nevertheless, in certain localities the macaques were experiencing a population growth rate of between 14 to 62%, with a sex ratio constantly in favour of males (Ménard *et al.*, 1986). Moreover the proportion of immatures was approximately 50% (Ménard, 2002). Unfortunately long term social and political conflicts prevented further research on the distribution of Algerian Macaque populations until recently. New data from the National Park of Gouraya (Grande Kabylie), show that the population of macaques is apparently increasing, being currently composed of six groups, averaging 27 animals per group and with a density of 17 individuals per square km (Mousli, 2003). An increase has also occurred in the national park of Chrèa, where the population is divided into four groups. Finally, monkeys have recently also been observed in Oued El Alleug (Chiffa), which is well outside the former distribution for the species (http://www.sos-magots.com). Thus, the Barbary macaque is probably less threatened in Algeria compared to Morocco probably due to the lower degree of human pressure on the habitat in this country. In fact in Algeria, researchers currently consider the animal to be in normal balance with its environment (Dehale and Chakali, personal communication, 2005), and it appears that overall, the population is stable or even increasing. Significantly, it is possible, perhaps even likely, that the reduction or absence of shepherds and their herds from the mountains as a result of the recent conflicts in Algeria have favoured the population of Barbary macaques in this country. Further scientific data however, are urgently needed to correctly estimate the impact of the human conflict on macaque habitat and its consequences for the demography of these animals in Algeria.

Principal causes of the decline of the populations of the Barbary macaque

The causes of fragmentation and reduction of the Barbary macaque populations are multiple. The principal one is the impact of Man on the habitat and ultimately on biodiversity within the species range. This is augmented by the fact that drought and the shortening of periods of snowfall in the Middle Atlas (Morocco) have allowed shepherds and others to remain in the forest for longer periods, even throughout the winter, when traditionally they have moved to lower lying areas. These people have further modified the water sources making them inaccessible to wild fauna and/or have started to camp in their vicinity. Human occupation of habitat within the macaque range has therefore become permanent over the last 10 to15 years and this has severely exhausted resources in a short space of time. Despite the remarkable capacity for adaptation of the Barbary macaque, the availability and diversity of food in its habitat are of primary importance. (Ménard *et al.*, 1985). Added to this is the increasing density of the livestock and the proliferation of mixed herds of goats and sheep at the expense of the traditional flocks exclusively of sheep. In particular, mixed herds exploit habitats more aggressively in high altitudes where monkeys still find refuge and some food. Drucker (1984) in fact demonstrated that livestock exploits forest resources more intensively than the populations of monkeys. The continuous overexploitation of the land removes most of the undergrowth, as is clearly evident in the degradation of the cedar forests of the Middle Atlas and the mixed forests of the Rif, and also has a harmful effect on wild fauna in general and on the macaque in particular by decreasing the amount of natural food available. Since the macaques are very sensitive to this intense exploitation of the habitat by domestic species, it has resulted in a progressive reduction in their density. The reduction in food resources on the other hand has resulted in macaques being more regularly present on the roads in the Middle Atlas where they beg for food* and water (Lavieren, 2005), producing the illusion that the population is growing, simply because it has become more visible, while the reverse is in fact the case.

In addition to mismanagement of the environment by Man as a cause of the decline in the species, the illegal trade in macaques is also having a significant negative influence. Infants are captured throughout Morocco and sold to tourists or to Moroccans resident in Europe. The new owner, however, usually gets rid of the animal when it becomes adult and once its natural wild and usually aggressive behaviour appears. More than 300 infant monkeys are captured and sold in Europe each year (Lavieren, 2004). This figure represents approximately 50% of the whole infant production of the combined Moroccan population, and largely exceeds the rate of off take in wild macaques which can be sustained (20%; Robinson and Redford, 1991 in Lavieren, 2004).

Hunting is yet another important factor contributing to the decline of the populations of primates (Oates, 1996 in Lavieren, 2004). Predatory dogs belonging to shepherds add to the extensive persecution of the macaques. Dog predation in fact accounts for the disappearance of a great number of young monkeys in the Middle Atlas. We noted that approximately 80% of the infants counted in 1994, disappeared between the summer and the autumn (Camperio Ciani *et al.*, 1999; 2004).

Due to the effect of habitat degradation, adult males are forced to constantly migrate in search of appropriate habitats and converge toward refuge habitat were foods still abound. In these areas male density increases and consequently the likelihood of conflict accentuates as does competition within and between troops for nutrients and sexual resources. (Camperio Ciani *et al.*, 1996, Camperio Ciani and Castillo 2000). All these factors make Barbary macaques more stressed in their usual habitat. Escalated fights and even adult conspecific killings have been observed and attributed to competition for resources and human habitat degradation. (Camperio Ciani and Marcharias, 2003).

---

* This is now particularly obvious at cedar Gouraud (Middle Atlas) where monkeys come close to visitors even pulling their clothes (in a behaviour which is similar to that seen with tourists in Gibraltar), but did not occur at all as recently as 1995. A similar situation has also recently developed in the surroundings of Aïn Kharzouza (locality overlooking the town of Azrou).

According to Azeroual (1995, not published; see Table 2), human encroachment into the natural habitat and increasing competition for resources is also causing the splitting of the groups in the Middle Atltas central area (Ain Kahla and El Kissarit), resulting in the reduction of the number of individuals but an increase in the number of troops of smaller size.

Table 2. Numbers of the macaques in the forest of Sidi M'guild (Middle Atlas: Morocco) (Azeroual, 1995 not published) (Col: colonies; Ind: individuals)

	1993		*1995*		*Change*	
	Col	*Ind*	*Col*	*Ind*	*Col*	*Ind*
Aïn Kahla: surface of 3708 ha	17	960	18	710	+1	-26
Kissarit: surface of 942 ha	6	525	9	910	+3	+73

Management of the populations of *M. sylvanus* in Morocco

Forest managers express the opinion that the population of Barbary macaques is currently increasing in Morocco and that this has caused bark stripping of the cedar trees in the Middle Atlas. They have therefore judged that population management is a necessity. In contrast, research shows that the size of macaque populations has dropped and that the animals strip the bark of the trees only to satisfy their requirements for water (Camperio *et al.*, 2001) and/or to obtain certain nutrients absent in the modified (impoverished) habitat (Deag, 1974; Drucker, 1984; Ménard and Qarro, 1999). Often impoverishment of the habitat results from hasty and inappropriate forest management actions, such as cutting of the green oak with the aim of facilitating cedar growth (Benabid, 2002) and large cedar logging followed by mostly ineffective regeneration plans. Oaks are also cut by companies for charcoal production and farmers and shepherds are allowed to enter the forest to cut tree branches for sheep and goat fodder.

Public management of populations of the macaque in Morocco during the last twenty years has seen measures that are both in favour of the animal, and to its detriment. The decrease in density, reported by Taub (1974), of 40 individuals per square km as against 60 to 70 individuals per square km four years before in Aïn Kahla (Deag and Crook, 1971), encouraged the then Moroccan Government to forbid the exportation of macaques and even to introduce some from Rocamadour in France*. In 1995, the *Administration*

*195 monkeys were introduced in 1980 in the Middle Atlas: areas of Aïn Leuh (45 individuals) and of Timahdite (150 individuals) (Haffane, 1981). Nobody knows the fate of these animals reared under artificial conditions and introduced into an area where the wild population is already in difficulty.

des Eaux et Forêts et de la Conservation des Soles, Direction de la Conservation des Ressources Forestières proposed a programme aimed at transferring groups of macaques to diminish their supposed pressure in the Middle Atlas. Groups should have been removed from the Middle Atlas and transferred to other Moroccan sites, from where this species had already disappeared, preferably corresponding to those designated as Sites of Biological and Ecological Interest (SIBE, 1996)**. The transfer of macaques was considered by Benabid (2002), to be a contribution to the destruction of the biodiversity in the National Park of Ifrane, although it had been created with the opposite intention. We cannot confirm that any transfer has taken place, although we did see two large cages approximately 2mx3mx3m equipped with sliding doors, with the wire showing evidence of monkey bites, in the high macaque density region of Ain Khala between 2000 and 2002. We strongly suspect that some captures were done with those cages, but that they were soon abandoned because the animals quickly learned to avoid them. From our enquiries to foresters we found that monkeys were also captured with guns and possibly anaesthetics. We do not know how many animals might have been captured and how many survived, but 40 animals were transferred to the National Zoological Park of Temara (near Rabat) (Haddane, personal communication, 2003).

In 2003, a formal request from *Le Haut Commissariat aux Eaux et Forêts et à la Lutte Contre la Désertification* was sent to the World Conservation Union (IUCN) asking for assistance in finding a solution to the alleged monkey-induced damage in the Middle Atlas, and both the authors' institutions offered help, by way of consultancy, forester training and seminars in collaboration with WWF-Med programme. In October 2003 a meeting was organised in Azrou with members of WWF-Med programme, and *Le Haut Commissariat aux Eaux et Forêts et à la Lutte Contre la Désertification* directed at forest officials, technicians and representatives of the Morocco nature conservation NGOs and the local community. A presentation on forest and macaque ecology was held, and the opinion of all the stakeholders attending the meeting on the causes of forest destruction was sampled. It emerged that all considered macaque over-population to be responsible for forest destruction. A field visit to the forest was organised for the second day, showing the impact of goats and sheep in the forest. The monkeys were censused both along roads and inside the forest along various transects. This was followed by a focus group discussion on the cause of forest destruction,

---

**11 sites were selected, by a commission, to receive the transferred monkeys, with the aim of distributing the population of the species beyond its current limits in the Middle Atlas. After visiting the 11 sites, the authors (Mouna and Camperio Ciani, not published), chose 4 areas which could receive groups of macaques. These areas are by order of decreasing importance: Jbel Bouhachem, forest of Aghbar, Jbel Tazerkount and Jbel Beni Snassen.

with the result that all members of the NGOs and the representatives of the local community (23 persons in all) radically changed their minds and stated that monkeys cannot be responsible for the degradation witnessed. Of the forestry officials (16) eight refused to answer and eight admitted that monkeys cannot be solely responsible. Although it had become clear to the NGO's and community representatives that monkey density is extremely low, foresters continued to state that monkeys "swarm" in the area.

Some recommendations and suggestions for conserving the Barbary macaque in the wild

General recommendations

Our recommendations, derived from this survey, range from short to long term goals as follows:

- Upgrade *M. sylvanus* from appendix II to appendix I of the Convention on International Trade in Endangered Species (CITES) and consider it as a threatened species according to IUCN criteria.
- Ban the illegal trade in monkeys in Morocco and enforce the ban by procecuting offenders.
- Prohibit the feeding of monkeys by tourists.
- Reduce the problem of roaming dogs, including their capture.
- Promote collaboration and exchange of information between researchers, managers and all environment stakeholders.
- Fence selected areas off for the exclusive use of wild fauna in Morocco and Algeria.
- Promote further studies to monitor the population size of the macaque in the two countries.
- Develop awareness in the human population of the region of the importance of the forest and its biodiversity and the danger of the illegal trade to wild fauna by:
    - Education programmes for children, including toys, cartoons, etc.
    - Involvement of stakeholders through questionnaires, participatory approach, interviews, media coverage.
- Develop ecotourism with the aim of reducing the pressure on the forest

and on the natural environment in general and converting it into an alternative financial resource for the local communities.

- Install corridors for the regeneration of the vegetation to ensure gene flow between the currently isolated populations of macaques and of other wild animals.
- Adopt an ecosystem approach to forest management as a whole for preserving biodiversity as a better option than single species conservation.
- Re-introduce new populations of macaques into areas from which they have disappeared, in particular those where minimum conflict with humans is predicted.

Recommendations for the Moroccan Middle Atlas population

- Allow monkeys access to water through the installation of suitable apparatus and by preventing shepherds from camping around water points.
- Progressively reduce the recent problem of mixed herds in the Middle Atlas by removing goats.
- Actively investigate a solution to the problem of the ownership of the herds which exhaust the resources needed by the inhabitants of the Middle Atlas.

Conclusions

Macaca sylvanus is the only species of monkey north of the Sahara and the only African species of macaque, all others being Asian. *M. sylvanus* is an invaluable component of forest biodiversity in North Africa where it has existed for thousands of years. More widespread in the past, the Barbary macaque is currently confined, in the wild, to Morocco and Algeria in specific forests and steep rocky areas where it is threatened by a number of factors. The current small sparse and unbalanced population structure of the species in the wild is the result of excessive human activity in its habitat and of the illegal trade and hunting. In North Africa in general, this environment is no longer as favourable a habitat for the Barbary macaque as certain authors formerly believed. If the current situation persists the animal will completely disappear from the area in the coming years, making it necessary that conservation efforts focus on national measures, conventions and international agreements. Educational programmes, also, are desirable to make people aware of the value and interest of the macaques at local and regional levels. In particular the beneficial role of the Barbary macaque in the ecosystem and its benefit to eco-tourism need to be stressed. Moreover a compromise should be found with the stakeholders for a sustainable exploitation, in particular of forest resources, by looking for alternative land uses followed by further management of the environment where *Macaca sylvanus* still finds refuge. Naturally one of the needs is more data, but we want to stress that active conservation must not await the collection of more data – we believe that we know enough now to know what needs to be done.

References

Benabid A (2002) Le Rif et le Moyen-Atlas (Maroc): Biodiversité, menaces, préservation. African Mountains High Summit Conference. Nairobi, Kenya 6-10 Mai 2002

Camperio Ciani A (1986) La *Macaca sylvanus* in Marocco supravvivenza o estinzione. Osservazioni personali e dati storico-demografici *Anthropologia Contemporanea* **9** 117-132

Camperio Ciani A and Castillo P (2000) The desertification process in the last natural forest of the southern Mediterranean region. In *Mediterranean Desertification research results and policy implications* pp. 471-480 Eds P Balabanis, D Peter, A Ghazi, e M. Tsogas. European Commission Div XII Research and Development, Vol. 2

Camperio Ciani A and Marcharias J (2003) Frequency of aggressive behavior and a case of mortal attach in wild *Macaca sylvanus* in the Middle Atlas region of Morocco *Human Evolution* **18** 3-4 123-130

Camperio Ciani A, Martinoli L, Capiluppi C, Arahou M and Mouna M (2001) Effects of water availability and habitat quality on bark-stripping behaviour in Barbary Macaques. *Conservation Biology* **15** 259-265

Camperio Ciani A, Arahou M and Mouna M (1996) *Macaca sylvanus* as biological indicator to monitor the cedar forest of Morocco Folia Primatologica **67** 63-80

Camperio Ciani A, Mouna M and Arahou M (1999) *Macaca sylvanus* as a biological indicator of the cedar forest quality. *Proceedings of the first international conference on biodiversity and natural resources preservation* pp 91-98, Ifrane

Camperio Ciani A, Palentini L, Arahou M, Martinoli L, Capiluppi C and Mouna M (2004) Population decline of *Macaca sylvanus* in the Middle Atlas of Morocco *Biological Conservation* **121** 4 635-641

Camperio Ciani A, Palentini L, Arahou M, Martinoli L, Capiluppi C and Mouna M (2005) Population decline of *Macaca sylvanus* in the Middle Atlas of Morocco *Biological Conservation* **121** 4 635-641

Cuzin F (2003) Les grands mammifères du Maroc méridional (Haut Atlas, Anti Atlas et Sahara): Distribution, écologie et conservation. Thèse de Doctorat (Ecologie Animale). Ecole Pratique des Hautes Etudes Sciences de la Vie et de la Terre, Université Montpellier II pp 1-351

Delson L (1980) Fossil Macaques, phyletic relationship and scenario of deployment in the macaques. In *The Macaques: Studies in Ecology, Behavior and Evolution* pp 10-32 Ed DG Linburg. Van Nostrand Reinhold, London

Deag JM (1974) A study of the social behaviour and ecology of the wild Barbary Macaque *Macaca sylvanus* L. Ph. D. thesis. University of Bristol.

Deag JM and Crook JH (1971) Social behaviour and "Agonistic buffering" in the wild Barbary macaque *Macaca sylvanus L. Folia Primatologica* **15** 183-200

Drucker GR (1984) The feeding ecology of the Barbary Macaque and cedar forest conservation in the Moroccan Moyen Atlas. In *The Barbary Macaque* pp 135-164 Ed JE Fa. Plenum Press, New York

Fa JE, Taub DM, Ménard N and Stewart PJ (1984) The distribution and current status of the Barbary Macaque in North Africa. In *The Barbary Macaque* pp 79-111 Ed JE Fa. Plenum Press, New York

Haffane M (1981) Le Macaque de Barbarie. Document du Laboratoire de Zoologie de l'Institut Agronomique et Vétérinaire Hassan II pp 1-13.

Lavieren Van E (2004) The illegal trade in the Moroccan Barbary macaque (*Macaca sylvanus*) and the impact on the wild population. Thesis MSc Primate conservation pp 1-62. Oxford Brookes University

Lavieren Van E (2005) Status of the Barbary macaque (*Macaca sylvanus*) population in the cedar forest, Middle Atlas Mountains, Morocco, 2005. AAP, Sanctuary for exotic animals Almere, The Netherlands: 1-25

Mehlman PT (1984) Aspects of the Ecology and Conservation of the Barbary Macaque in the Fir Forest Habitat of the Moroccan Rif Mountains. In *The Barbary Macaque* pp 165-199 Ed JE Fa. Plenum Press, New York

Ménard N (2002) Ecological plasticity of Barbary Macaques (*Macaca sylvanus*). Primate Behavior, Ecology, and Conservation *Evolutionary Anthropology, Suppl.* 1 95-100

Ménard N and Qarro M (1999) Bark stripping and water availability: a comparative study between Maroccan and Algerian Barbary Macaques (*Macaca sylvanus*). *Rev. Ecol.* (Terre Vie) **54** 123-132

Ménard N and Vallet D (1993) - Population dynamics of *Macaca sylvanus* in Algeri a: An 8 – year study *American Journal of Primatology* **30** 101-118

Ménard N, Vallet, D and Gautier-Hion A (1985) Démographie et reproduction de *Macaca sylvanus* dans différents habitats en Algérie. *Folia Primatologica* **44** 65-81

Ménard N, Amroun M, Mohamed-Said R and Gautier-Hion A (1986) Status of the Barbary macaque (*Macaca sylvanus*) in Tikjda forest, Algéria. *Primate Conservation* **7** 35-37

Ménard N, Hecham R, Vallet D, Chikhi H and Gautier-Hion A (1990) Grouping Patterns of a Mountain Population of *Macaca sylvanus* in Algeria – A Fission-Fusion System? *Folia Primatologica* **55** 166-175

Mouna M, Arahou M and Camperio Ciani A (1999) A propos des populations du singe magot (*Macaca sylvanus* L) dans le Moyen Atlas. *Proceedings of the first international conference on biodiversity and natural resources preservation* pp 105-109, Ifrane

Mousli ML (2003) Ecologie des populations du Magot (*Macaca sylvanus* L.), dans le parc national de Gouraya (Kabylie, Algérie). Impact des changements climatiques sur l'écologie des espèces animales, la santé et la population humaine maghrébine. Journées Scientifiques organisées par l'Association Groupe de Recherche pour l'Environnement Urban et Rural, Rabat, les 9, 10, et 11 juillet.

Oates JF (1996) Habitat alteration, hunting and the conservation of folivorous primates in African forests *Aust. J. Ecol* **21** 1-9

Robinson JG and Redford KH (1991) Sustainable Harvest of Neotropical Forest Mammals. In *Neotropical Wildlife Use and Conservation* pp 415-429 Eds JG Robinson and KH Redford. University of Chicago Press, Chicago

Scheffrahn W, Menard N, Vallet D and Gaci B (1993) Ecology, demography, and population genetics of Barbary macaques in Algeria *Primates* **34** 3 381-394

Sehhar E (2003) Mammifères terrestres, Jbel Gourougo. Rapport Med Wet Coast Maroc. Ministère de l'Aménagement du Territoire de l'Eau et de l'Environnement et le Département des Eaux et Forêts et de la Lutte contre la Désertification pp 1-37

Sehhar E (2004) Diagnostic en Mammalogie (SIBE de Tamga et du Jbel Moussa). Rapport Projet GEF de gestion des Aires Protégées TF – 023494 – MOR, Septembre 2004

SIBE (1996) Etude des Aires Protégées. Plan Directeur des Aires Protégées au Maroc. Volume 2. Les Sites d'Intérêt Biologique et Ecologique du domaine continental. MAMVA. Administration des Eaux et Forêts et de la Conservation des Sols. BCEOM – SECA, BAD EPHE. ISR. IB pp 1-166

Taub DM (1974) *A report on the distribution of the Barbary macaque Macaca sylvanus in Morocco* pp 1-45 Department of Anthropology, University of California Davis, California

Taub DM (1977) Geographic distribution and habitat diversity of the Barbary Macaque *Macaca sylvanus L. Folia Primatologica* **27** 108-133

Taub DM (1978) Aspects of the biology of the wild Barbary Macaque (Primates, Cercopithecinae, *Macaca sylvanus* L. 1758): Biogeography the mating system and male-infant associations. D. Phil. Thesis. Univesity of California, Davis.

Taub DM (1984) A brief historical account of the recent decline in geographic distribution of the Barbary Macaque in North Africa. In *The Barbary Macaque* pp 71-78 Ed JE Fa. Plenum Press, New York

Waters SA, Fa JE, Akissou M, El Harrad H and Hobbelink ME (2005) Holding on Djebela : Barbary macaques in Northern Morocco. *Oryx* (in Press)

Human and environmental causes of the rapid decline of *Macaca sylvanus* in the Middle Atlas of Morocco

A Camperio Ciani[1], and M Mouna[2]
[1]*Dipartimento di Psicologia Generale, Universita' Degli Studi di Padova, Via Belzoni, 80 - 35131 Padova, Italy*
[2]*Département de Zoologie et d'Ecologie Animal, Institut Scientifique, B. P. 703 Rabat 10106, Morocco*

Overview of *Macaca sylvanus* in the wild

The genus *Macaca* is the most widespread genus of non human primates in the world. Its recent distribution stretches from northern Africa to Japan, from South Korea to the Sunda archipelago, and most of continental Asia, including the Indian subcontinent (Camperio Ciani, 1986a). Macaques can be found in tropical and montane forests as well as plains and deserted areas, and many populations have also adapted to an urban environment. Primarily due to their functional social organisation and varied feeding ecology (Camperio Ciani, 1986a), macaques probably represent the most flexible and adaptive of all non-human primates. In particular, Macaca *sylvanus* is well known for its behavioural flexibility and ecological adaptability. However, throughout its whole phylogenetic history, from the Pleistocene to the present, the Barbary macaque has progressively disappeared from all the circum-Mediterranean habitats, except from Morocco and Algeria (Camperio Ciani, 1986; Thirgood, 1984). Even in these two countries, it lives in a relatively endangered condition (Fa *et al.* 1984). At present, we estimate, that the total living population is no more than 10,000 individuals, distributed between both countries (Camperio Ciani *et al.* 2004). Moreover, the high degree of fragmentation and restricted gene flow between sub-populations (Sheffrahn *et al.* 1993; Sheffrahn, 1999), makes the effective population even smaller and more vulnerable.

The only exception is in the largest remaining mixed oak and cedar forest of the Middle Atlas, where this species has been described as abundant and relatively undisturbed and where around 70% of the total remaining wild *Macaca sylvanus* live (Camperio Ciani *et al.* 2004 Mouna *et al.* 2000). Even here, however, the situation has deteriorated considerably and the Middle Atlas mixed forest region has now also become a vulnerable ecosystem in which the population of *Macaca sylvanus* is rapidly declining. Here we will try to describe this decline and demonstrate that it is entirely attributable to human factors, which, through their adverse impact on the habitat are having dramatic consequences for the demography of these last wild populations. Although this chapter focuses on the Middle Atlas population, we suggest that the findings are relevant to the species as a whole.

Middle Atlas ecology

The Middle Atlas forest ecosystem is characterised by cedar *Cedrus atlantica* and various species of oak including *Quercus rotundifolia* and *Q. faginea*. The only other mixed cedar forests can be found in small and relict areas of Lebanon, on the island of Cyprus, and in the south west Anatolia region of Turkey. The Middle Atlas mosaic forest system is, therefore, the last remaining large north African forest. Here the forest tree composition can reach 40 different species, and up to 300 varieties of herbs and shrubs have been recorded (Drucker, 1984). In addition to mammals, a whole range of endemic birds, reptiles and invertebrates helps make these forests a unique and still largely unknown reservoir of biodiversity. It was in this forest ecosystem that the last populations of the Berber lion and leopard were seen. The first became extinct in the 1940s, although the latter is still present in Azilal, a remote region in the south-east of the Middle Atlas. These forests are also still host to small populations of lynx (*Lynx serval*), and possibly hyena (*Hyaena hyaena*) as well as a relict population of Berber deer (*Cervus elaphus barbarus*). Foxes (*Vulpes vulpes*), jackals (*Canis aureus*), and wild boars (*Sus scrofa)* can still be found with relative abundance, and the forest is home to the last large population of Barbary macaques.

Over the last three millennia, the Middle Atlas high land forest ecosystem has been substantially modified by Man (Thirgood 1984), especially the nomadic Berber tribes who would bring their flocks of sheep and goats to pasture here in the summer. The land, in fact, is subdivided into territories with traditional exclusive use by specific Berber tribes. These tribes used this pastureland during the summer according to strict rules of family priorities and friendships. The cedar wood has been used since French colonial times as a source of valuable timber, while the oak has been used for both construction and as fire wood. Minor traditional activities included charcoal production, lichen collection for the cosmetic industry, cedar oil production and ethno-medicine. The forests of this region are also extremely important, and have always been used, as a water reservoir for the whole agricultural system of the Moroccan plains.

Long term field research

We first visited the Middle Atlas in 1984, in order to study macaque ecology and behaviour. As a result of the progressive deterioration of the habitat in the area of our study, we decided in 1992 to initiate a project to monitor the forest ecosystem and its wildlife. We identified a 483 km^2 mosaic forest ecosystem in the central Atlas region, east of Ifrane, Azrou and Ain Leuh (33°15'N, 5°15'W) and conducted field sessions with colleagues from the Scientific Institute of Rabat, along with previously trained graduate and undergraduate students. With the help of a GPS, we identified 16 rectilinear segments across the various forest ecosystems, varying in length between 4 and 7 linear km, totaling a 93.5 km circuit within the forest (Figure 1). This virtual circuit, made up of the sixteen transect segments, was then periodically walked thirty times between June 1994 and October 2002 (see detailed methods in Camperio Ciani *et al.*, 2001 and 2005).

All indicators, relating to habitat, human presence and activity, presence of domestic animal flocks, primates and wildlife, recorded along the transect segments, were selected as a compromise between the effectiveness of describing transformation and impact of the habitat on the one hand and the possibility of effective data collection and replication by teams with different levels of experience, on the other. In an effort to increase statistical power and to render our results comparable with data from previous studies (Deag, 1974;Taub, 1977), we further classified our data into three expedition blocks of 12, 10, and 8 circuits. The first 12 circuits refer to the first 4 expeditions occurring between 1994 and 1995, the next 10 circuits refer to 3 expeditions that took place between 1998 and 2000, and the last 8 circuits refer to the single expedition of 2002.

Transect versus focal troop method

The transect method, which we used extensively, is neither the only method to study demography in the wild, nor the most accurate. We chose it because it is the most efficient and describes large areas most realibly. The focal troop sampling method, on the contrary, is an in depth study of a restricted group of animals, and can give very accurate results regarding age- sex-class structure, birth rate, and mortality rate and has been extensively used for wild *Macaca sylvanus* (eg. Deag, 1977 Mehlman, 1989 Menard &Vallet, 1993). Even local density of animals can be calculated using focal troop sampling, by following troops for a long time, and estimating 1) their home

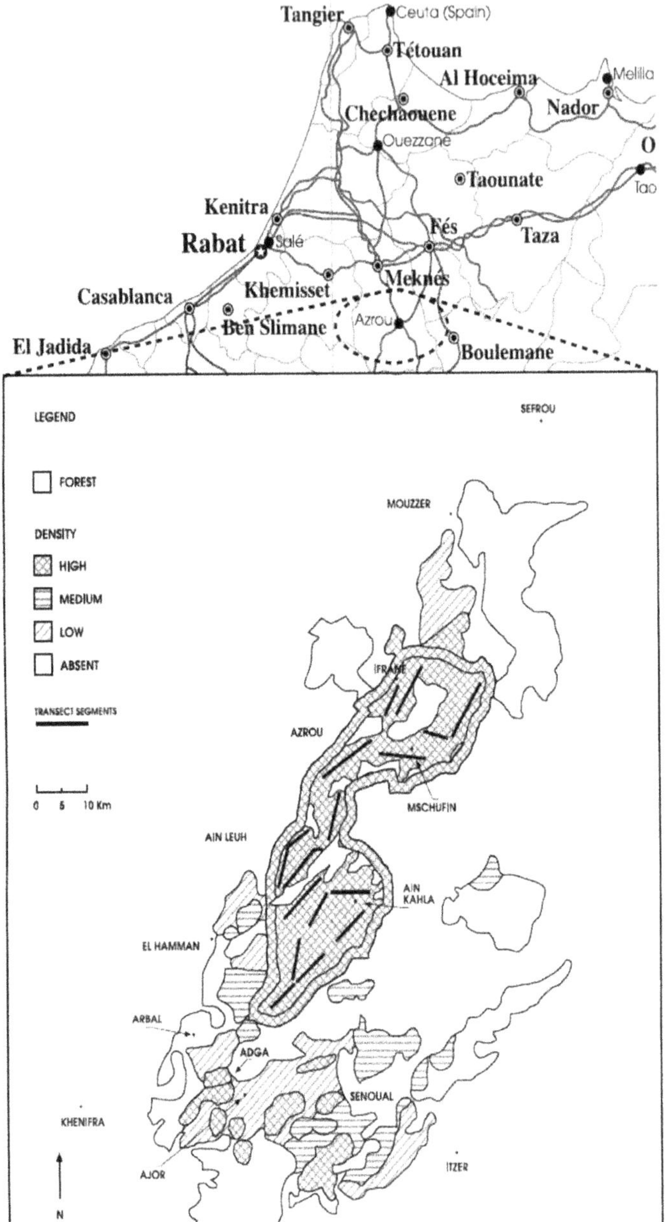

Figure 1: Geographical location of our study area in the central Middle Atlas, including the distribution of forest cover, macaque distribution as previously reported (Taub *et al.* 1984), its relative density, and the location of the 16 transect segments forming our monitoring circuit (Reprinted from Biological Conservation, 121,Camperio-Ciani, Palentini, Arahou, Martinoli, Capiluppi and Mouna, Population decline of *Macaca sylvanus* in the middle atlas of Morocco, 635-641, © 2005, with permission from Elsevier).

range, 2) the rate of overlap with neighbouring troops, and 3) size of the neighbouring troop (*eg*. Mehlman, 1989). The focal troop method, however, implies the individual recognition of all members of the troop, which takes extremely long to achieve in the field where there is a low density of animals. The focal troop method, after an adequate time in the field, can also be very precise for estimating demographic patterns and is thus more accurate than the transect method. However, it also has drawbacks. Most researchers using the focal troop method, in selecting their research site, first search for an adequate area in terms of feasibility and abundance of target animals (eg Deag, 1977) and this tends to overestimate average density. Therefore, the focal troop method, which has a very high internal validity, has very little external validity, namely outside the area where it has been conducted. The transect method is admittedly based on quick and sometimes occasional sightings, and does not require individual recognition of the animals encountered. If the environment is variable or mosaic like, as it has been extensively shown in the Middle Atlas (eg. Mouna *et al.*, 1999 ; Menard *et al.* 1999), then the results of the transect method become more realible on a large scale. In fact it has often been shown that animal density and demographic structure vary substantially according to habitat (see below). In conclusion, we could say that the focal troop method is an adequate choice for a local long term project, but may not be the best choice in terms of representativeness of a whole area. The transect method, on the other hand, is efficent in covering a large surface or region, passing across all habitats and environments. As this method allows an adequate number of replications, the researchers can always approach a better estimate of density, troop location, and age and sex class composition. Even home range can be estimated, by calculating the average distance between subsequent sightings along the transects. Another advantage is that, in not needing individual recognition, data collection can start considerably sooner, and with less training (therefore allowing the help of multiple assistants).

Density decline

Figure 2 shows how the relatively high number of replications (30) of our statistical unit (a 93.5-km-long circuit comprising 16 segments) helped reveal a trend indicating that in the ten years since the beginning of our investigation, *Macaca sylvanus* density for all age and sex classes and in all habitat types has suffered a serious decline. Density has fallen from an average of 28-30 individuals per km^2 observed in the expeditions conducted between 1994-1995, to 7-10 per km^2 observed in the 2002 expedition. Indeed, the regression analysis between densities per circuit and time elapsed from the beginning of the study, shown in Figure 2, is highly significant ($r = 0.729$, $p < 0.001$), and strongly suggests that the macaque population in this area is dramatically and continuously decreasing.

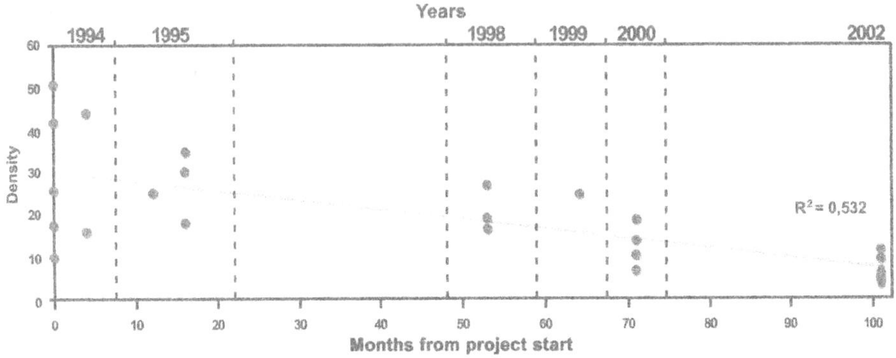

Figure 2. Regression analysis between time elapsed and the estimated macaque density in each of the 30 replications of the 93.5 km long circuit between 1994 and 2002. Each data point represents a single replication of average density per km² for the whole circuit, starting in 1994 and ending 100 months later in 2002. (Reprinted from Biological Conservation, 121, Camperio-Ciani, Palentini, Arahou, Martinoli, Capiluppi and Mouna, Population decline of *Macaca sylvanus* in the middle atlas of Morocco, 635-641, © 2005, with permission from Elsevier)

These data are even more dramatic when compared with the density estimate by Deag in 1968 (Deag 1974) of between 50 and 70 animals per km² in the central region of the Middle Atlas (Ain Khala) and Taub's more recent estimate of 44 individuals per km² (Taub,1984). This suggests that the population density of *Macaca sylvanus* in the largest surviving population has decreased by around 90% in less than 30 years.

Causes of density decline

Predators

Large cats such as leopards and lions, that naturally preyed upon macaques, are now extinct in the region. Other predators, including foxes and jackals, are still present however, and as effective predators of infant macaques, may be enough to limit population growth. During our study, we directly observed two predations by foxes and one by a jackal, but noteworthy were at least six further predations witnessed which were by domestic dogs. Domestic dogs have increased their presence in the forest following shepherds and flocks. They are usually medium to large sized dogs and are always found in groups of three or more. These dogs are stealthy and fast and are very effective in snatching infants from adult macaques.

Habitat degradation

Habitat degradation, both in terms of reduction in the extent of forest cover and in the deteriorating condition of the undergrowth, seems a much more likely cause of the progressive reduction in numbers of *Macaca sylvanus*. Figure 3 shows the proportions of seven habitat sub-types, for the three expedition blocks. We can see the progressive reduction in cover of both closed and open cedar forest and the increase of mixed open forest, open oak forest and meadows in the more recent years. Figure 4 shows that the density of macaques (as the number of individuals sighted in each square kilometer of each forest habitat sampled) has fallen dramatically. The total decline in animals in the area is therefore due to both reduction in the surface of forest habitat and in the density of macaques in each forest habitat.

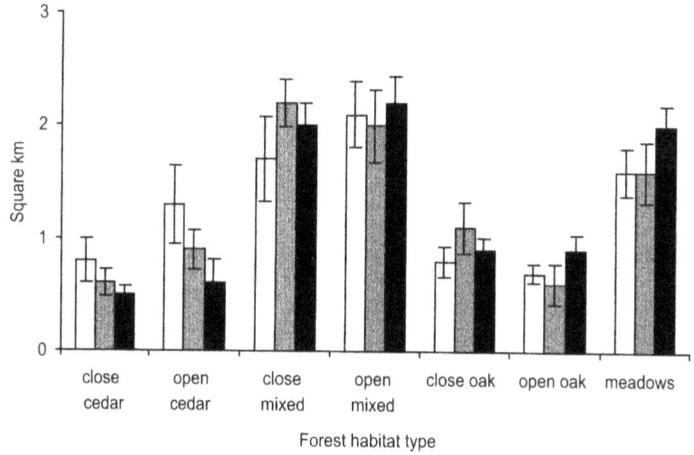

Figure 3. Forest surface, relative to sampled habitat. Data have been clustered into three blocks; columns refer to expeditions carried out in 1994-1995 (12 circuits) □, 1998-2000 (10 circuits)▩, 2002 (8 circuits)■. Bars indicate 95% confidence limits (data modified from Camperio Ciani *et al.* 2005).

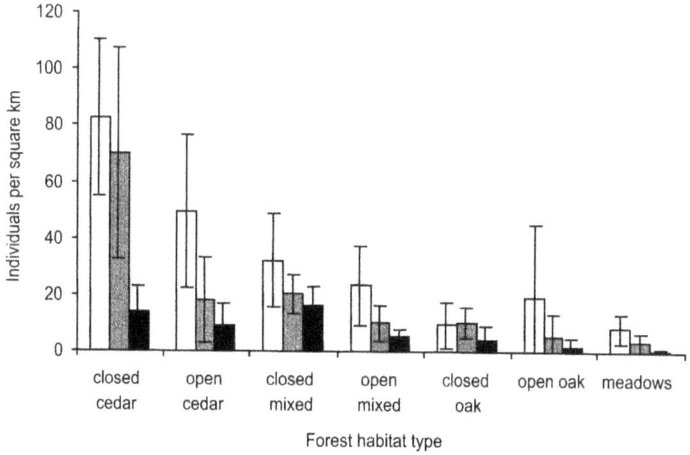

Figure 4. A comparison of the decline in number of Barbary macaques sighted in the same three expedition blocks, in each habitat sampled. Columns refer to expeditions carried out in 1994-1995 (12 circuits) □ , 1998-2000 (10 circuits) ▩, and 2002 (8 circuits) ■. Bars represent 95% confidence limits (data modified from Camperio Ciani *et al.* 2005).

Overgrazing and Human impact

The condition of the forest is best described not by amount of tree cover but by the condition of its undergrowth. Large cedar trees do not die easily, and adult oak trees can grow again if cut. On the other hand, if the undergrowth is damaged, tree regeneration is precluded, and sooner or later the forest is destined to die. Further, most nutrients for wildlife lie in the undergrowth, and the carrying capacity of the habitat, relative to macaques depends on the condition of this undergrowth.

The major threats to the undergrowth are:

1) soil erosion, which depends on forest cover and surface steepness, and

2) impact of grazing by domestic livestock.

Livestock grazing is particularly devastating when, as in this case, the habitat is used by a combination of sheep (90%) which graze the grass and goats (10%) which pull out roots from the soil and strip the bark from small trees and bushes. The maximum density of domestic animals grazing in the forest was estimated for sheep at 130/km^2, during the 1994-1995 expeditions, vs. the 316/km^2 observed for the 1998-2000 expeditions. The same trend is evident for goats, whose density went from 18/km^2 in 1994-1995 to 51/km^2 recorded between 1998-2000. The data from our last expedition (October 2002) indicate an apparent inversion of this trend, with a relative decline in the density of grazing animals. Unfortunately however, the data from the 2002 expedition are not comparable because they were collected relatively late in the season, when many flocks had already been relocated to winter in the plains.

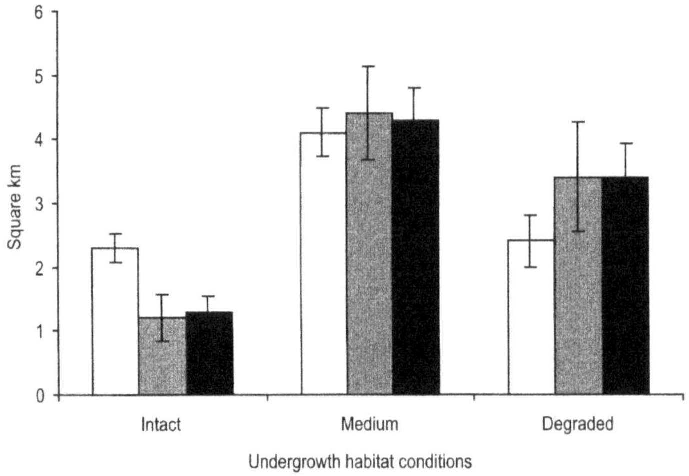

Figure 5. Average surface area (km^2) of forest undergrowth, according to its condition. Columns refer to expeditions carried out in 1994-1995 (12 circuits) □, 1998-2000 (10 circuits) ▩ and 2002 (8 circuits) ■. Bars represent 95% confidence limits (data modified from Camperio Ciani *et al.* 2005).

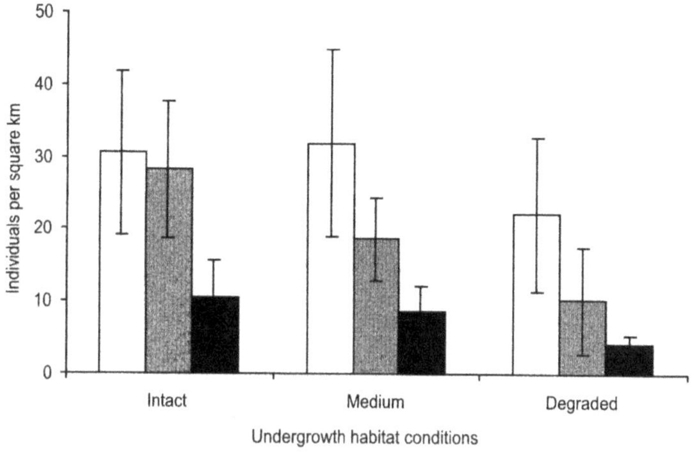

Figure 6. Relative density of macaques observed in each undergrowth habitat type. Columns refer to expeditions carried out in 1994-1995 (12 circuits □, 1998-2000 (10 circuits) ■ and 2002 (8 circuits) ■. Bars represent 95% confidence limits (data modified from Camperio Ciani *et al.* 2005).

Figure 5, shows the decline in the proportion of intact areas and the increase in degraded terrain between the first two blocks, whereas data from the 2002 expedition indicate that the situation has evidently stabilised over the last few years. Only 13% of the terrain surveyed within the forest could be classified as ***intact***, while 50% presented recent signs of grazing (***intermediate***), and 37% showed alarming signs of overgrazing (***degraded***). The decline in monkey density with respect to underbrush type (Figure 6) is severe for each of the three underbrush surface categories, and highly significant for intact and intermediate surfaces. Regeneration of young cedars, as an indicator of forest regeneration in general is extremely reduced. We witnessed whole areas in which no regeneration occurred at all. Most of the naturally regenerating cedars occur in oak, rather than cedar forest, or at the margin of the forest. Along our walked circuit and vehicle-surveyed area, we estimated a regeneration success, (defined as trees that had reached two meters in height), in no more than 10-15% of the the forestry department's artificial regeneration projects.

As mentioned earlier, Barbary macaques are considered by most researchers to be an important biological indicator of habitat status in this region. In fact we have shown that the age and sex classes vary predictably according to forest condition, from intact forests to degraded ones. Population demography can therefore be used to estimate forest condition (Camperio Ciani *et al.*, 1996; 1999, 2000; Mouna *et al.* 1999; Menard *et al.* 1999; Van Lavieren, 2004). The various macaque densities observed in the 3 habitat types in this study, suggest that macaques in this region mainly prefer cedar forests, especially closed canopy forests and that they avoid areas with degraded terrain.

When compared to those from previous studies conducted in the 1970's, our data underscore an alarming trend in this region in forest degradation, which we have been further witnessing in recent years.

Demographic trends of *Macaca sylvanus*

Our transect counts and consequent density estimates are not the only data suggesting that the macaque population is suffering. Two other studies on the *Macaca sylvanus* population in Morocco are pertinent to the present one for historical comparison. Deag conducted the first in 1968 in the Ain Khala region, within our own study area. Using the focal troop method, the study recorded a density of about 44-70 individuals per square km (Deag 1974). The second study (Taub, 1977), was performed as part of a general survey of all macaque habitats in Morocco and reported a density estimate for the central region of the Middle Atlas of 44 individuals per square km.

Table 1 shows the proportion of three age sex classes, and the density of Macaques in our own studies and those of the others for comparison. The first row reports the study of Deag (1973), the second the study by Mehlman in the marginal habitat of the Moroccan Rif. It can be clearly seen again that the process of density reduction is progressing at increasing speed; density is now approaching that in the degraded areas of the Rif. In the third column, we show a standard way to compare the proportion of immatures in the population (Southwick *et al.*, 1988). The authors estimate that in a Macaque population with a proportion of over 50% immature animals, the sex ratio corrected population can replace itself, whereas below 50% it will disappear. In the Middle Atlas, it can be seen that it had already dropped below this threshold once in 1995, and it is approaching it again in our last study in 2002. The proportion of immatures is a strong indicator of the future prospects of the population in terms of growth versus reduction (Camperio Ciani 1999 and Camperio Ciani *et al.*, 2005). In her focal study in the Ain Kahla region during 1995, Menard (1999) indicated a dramatically low proportion of immatures, with many groups below survival proportions.

Poachers, often witnessed during our study along the transects even during the day, are very effective in capturing infants during the tourist season by using dogs and long poles. Infant macaques are then offered to the tourists for €50-100 each (Camperio Ciani 1986, Van Lavieren, 2004). Human related habitat degradation and long summer droughts that have affected the region since the late 1980s have further taken their toll in killing infant macaques.

Table 1. Demographic comparison between studies in the Middle Atlas

Study	*Males**	*Females*	*Immatures*	*Imm.Agg.***	*Density****
Ain Kahla 1968	26,2	24,5	49,3	52,1	44
Gomara 1983	25,3	24,7	50	52,9	6,73
Our study 1995	36,8	23,1	40,1	42,1	28,19
Our study 2000	33,2	21,2	45,6	65,1	16,43
Our study 2002	31,6	27,1	41,3	50,3	7,24

* Values in percentage
** Percentage of immatures with a corrected male sex ratio of .5
*** Density/km^2, with focal troop method in the Ain Kahla and Gomara region, and with the transect method in our study.

Poaching, predation, and starvation are killing many infants every season and the adult population is not adequately replaced and progressively ages and declines. We suspect that this is happening also to the rest of the wildlife in the central Middle Atlas region and adequate monitoring is urgently required for these animals also.

As shown in column 2 of Table 1, there is a proportional excess of males in the area that, according to our biological indicator model (Camperio Ciani *et al.*, 1999; Camperio and Castillo 2000), indicates that this region is a refuge area, where males converge from even more degraded areas. We have shown in fact that a progressive increase of adult males in the population can be ascribed to immigration from neighbouring regions that are experiencing even worse habitat destruction than the area monitored (Camperio Ciani *et al.* 1999 Camperio Ciani and Castillo 2000). Adult male increase can have disruptive consequences on macaque population dynamics by increasing sexual conflicts and decreasing infant survival during the mating season (Paul and Thommen, 1984; Paul *et al.* 1996).

Conservation implications

Forest habitat reduction

Forest cover is progressively decreasing for two reasons: logging still occurs in oak forests, both for charcoal production and for fire wood. In the cedar forest, the decrease is due both to timber production without sufficient regeneration, and to illegal removal of branches by shepherds to feed their flocks in winter (see Figure 7). Branch breaking is particularly harmful to the forest: the trees are exposed to fungal attack through the exposed wood and soil loss increases dramatically due to erosion. Weakened de-branched trees are attacked by xilophagus insects, *Scolites* spp. which drill holes in the bark and feed on the cambium, and by a variety of wood fungi, which kill the remaining trees. Regenerating cedar trees are cut for fire, eaten by goats ranging in the forest, and eventually attacked by monkeys which strip their bark.

Since spring 2002, for causes as yet not fully established, the huge monumental cedars, the only class of trees relatively unharmed up to now, have also suddently begun to die. In a recent survey in 2003 we witnessed up to 90% of the monumental cedars had already died in certain areas, which means thousands of huge trees aged three hundred years or more.

From these data and other observations, we note how cedar forests (the habitat where *Macaca sylvanus* was most frequently sighted in this region in the early years of our study) have undergone a serious decline since the project's inception (Figure 3). Comparison of similar data collected by Drucker in the 1980's (Drucker, 1984) indicates that the ratio of open closed to closed canopy forest has decreased from 1.05:1 in 1980 to 0.36:1 in 2002.

The underbrush has not followed a better fate than the forest canopy; forest grazing has tripled in intensity over the past ten years and the undergrowth has become seriously degraded. Figure 5 shows how the underbrush in our sampled areas has also suffered substantial and progressive damage, mainly due to the growing impact of grazing by mixed flocks of goats and sheep.

The regeneration of cedar has been recorded mainly in oak forests and shows the strict interdependence between cedars and oaks in the region. In the Middle Atlas these two trees have an interdependent life cycle, cedars grow in the soil under oaks and eventually outgrow oaks; the latter will regenerate when large cedars eventually die and fall. At present, most oak forests are heavily grazed and natural regeneration occurs only occasionally. Assisted regeneration, which is the responsibility of the *Administration des Eaux et Forêts et de la Conservation des Soles (Eaux et Forêts)*, is doing rather poorly, for a number of reasons:

- most planting for regeneration takes place in open meadows where cedar grows with difficulty,
- cedar trees are planted at less than one year old, a very early age, and
- fencing is inadequate to prevent access by mixed flocks of sheep and goats for long enough, despite the efforts of rangers.

Human factors as causes of habitat degadation

The main causes of the recent decline of the central Middle Atlas region, and its *Macaca sylvanus* populations, are:

- human population growth,

- tribal sedentarisation,
- excessive exploitation of the wood in the forest for fodder, firewood and timber, and
- overgrazing of forest underbrush by mixed flocks of goats and sheep that further results in soil loss by erosion, and drastically reduces forest quality and regeneration.

Human population growth can be easily appreciated from the rapid urban development of the villages of Ifrane, Azrou, and Ain Leuh. Whole new neighbourhoods have appeared everywhere in the last ten years, and all these, once rural villages, have literally exploded. These new urban neighbourhoods make an increasing demand on water resources and consequently a whole new network of water pumping stations has appeared within the forest with yet undetermined consequences for the water table.

The traditional transhumance of the Berber tribes who drove their flocks out of the region toward the plains at the end of the summer, is now delayed and reduced. A growing proportion of Berber shepherds, using plastic covers to improve their tents, or building brick structures, remain in the forested highlands all year round, causing continuous pressure on the environment. This is evident, for example, around lake Affenourir, where around the now permanent Berber shepherd settlements, the forest is completely degraded and de-branched and the soil severely eroded.

A Berber settlement in the Middle Atlas, showing permanent structures that allow the shepherds to remain in the area throughout the winter. Copyright J. Cortes

Most people encountered in oak forests along the transects were spotted collecting firewood on the ground and cutting oak branches. Hundreds of donkeys and mules loaded with firewood can be encountered any day of the year. Charcoal production is managed in oak forest, but it takes its toll twice over, in fact, since once the charcoal producing activity is finished, regeneration in the area is inhibited by grazing, and continuous firewood collection in the cleared areas.

Logging in cedar forest is organised and managed by *Eaux et Forêts* and is based on:

1) an over optimistic estimated regeneration rate (80% instead of the actually recorded 15%),

2) a plot rotation reduced from 120 to 80 years, which means that each plot is cut only after 80, not 120 years as previously, with consequent reduction in forest regeneration.

If current practices continue, the logged forest will never be replaced.

As discussed above, the most important impact of all is that of the continuous growth of flocks pastured by the Berbers. This growth in the amount of livestock contrasts with the extremely poor living condition of the local Berber population. In fact, we discovered that most of the sheep grazing in the region do not belong to the local Berber tribesmen. They belong to wealthy urban investors, usually Arabs, who use their revenues to buy large flocks and allocate them for pasture to the local tribesmen under disadvantageous renting conditions (Schembri, 2003). The fact that flocks belong to wealthy and powerful people increases the potential for corruption of the often underpaid forest rangers. We found that for €100 it is possible to cut branches in a forest plot for one month, with the complicity of a ranger, while fodder for the same period would cost ten times more in the market (Schembri, 2003). The shepherds get paid with fifty of the lambs born each year, but only if the birth rate excedes one lamb for each of one third of the females present in the flock. It can be understood that these people make very little profit from these renting contracts. In bad years, when few lambs survive, there may be no profit at all. As a consequence they need to keep a few goats for their own survival needs. The presence of goats, as we have seen, dramatically increases the damage to the habitat. Berbers have traditional rights to use the forest for their livestock and at traditional levels, the grazing impact of these rights was sustainable for many years. But by accepting rent contracts from external investors, they are selling these rights below cost to others that are not compensating for habitat destruction, and yet gain most of the profit. Because of the rights of the Berber tribes, the National Forest Authority cannot fence off more than 20-30% of the forest land at one time, hence further reducing potential for natural and assisted forest regeneration. Even the areas protected with wire fences are not sufficiently secure and occasional entry into these areas by mixed flocks may cause irreversible damage.

Macaque impact on forest conservation

Within this overall picture, it is also true that macaques appear now to be causing some economic damage to the forest. By stripping bark from cedars, they damage timber production but kill very few trees (Menard *et al.* 1999). We showed that this habit is due to the absence of access to water during the dry season (Camperio Ciani *et al.* 2001) and has a cyclical variation depending on summer precipitation. The behavior is indirecty triggered by local people who permanently occupy water sources for their livestock, with people and dogs preventing access by wildlife to water. Other researchers have suggested that lack of nutrients in the macaques' diet is a cause of bark stripping (Menard *et al.*, 1999). The scale of the impact of bark stripping by monkeys in the ecosystem is minimal however and of little relevance when compared to human-generated impacts. We believe that given this global picture, they are a marginal problem (Camperio Ciani *et al.*, 2001). However, if something has to be done, the action should take the form of pilot projects to study feeding ecology and to allow macaques to regain access to water. These measures should have priority over plans to reduce their density as has been repeatedly suggested by local forest authorities.

We believe that we have exhaustively shown that macaques are declining rather than growing in numbers, and that bark stripping is not a result of increased density (Camperio Ciani *et al.*, 2001).

On the contrary, the macaques should be considered as adding richness to the forest. With their feeding activity they disperse seeds, promote regeneration, control insects, and have an important perspective as an eco-tourism attraction. Macaques are not the culprits of forest degradation but rather the victims and as an integral part of the natural eco-system should be protected.

Recommendations

Forest ecosystem restoration in this scenario will be a difficult and complicated task, but necessary if we want to protect and develop the last remaining large north African forested area and its unique animal and plant biodiversity. To accomplish this task it is first urgently necessary to arrest or reduce the sources of the impacts that are currently destroying the forest, its biodiversity, its wildlife, and its water resource potential.

The first step is to reduce current unsustainable Man made impacts, by a series of actions:

1) *Promote natural regeneration* implementing and enforcing a better forest protection plan, protecting not only cedar and mixed forests, but oak forest as well which is where most of the natural cedar regeneration occurs. Assisted regeneration techniques should be varied by planting within the oak forests, at least four year old cedars selected for drought resistance, as has long been done successfully in Turkey (Ishik, pers. comm).

2) *Intervene with legislation* and effective follow up action to prevent the external flock renting strategy and consequently to discourage the local tribesmen from keeping a combination of goats and sheep in their flocks.

3) *Subscribe to a temporary, but absolute, moratorium* on logging activity, at least until the recent epidemic of the mature cedar trees is better understood and resolved; these forests should be considered at present as a fundamental fresh water reservoir and not also as a source of income through logging.

4) *Improve the access of wildlife to water as a method of combatting bark stripping*, and commence a focused study on the feeding ecology of the macaques.

5) *Continue and expand the monitoring of the environment* to measure the impact of all possible actions on the habitat.

At the local population level it is necessary,

1) With young people:, to *start pilot awareness projects* in the schools of the area, through booklets, posters, encounters, on the role of the forest in the well being of their future, and on the risk of desertification.

2) With adults: *the local tribal communities must be involved in identifying all possible solutions*, using a complete set of available techniques such as focus groups, human dimension, participatory encounters, *etc*. Potential conflicts between all habitat stakeholders must be urgently identified, including the Water and Forestry Department, tourists, shepherds, farmers and all forest workers. The goal is to agree on a set of viable projects at lower habitat impact than present ones.

Following preliminary interviews conducted by our team in the area, between 1999 and the present, a preliminary list of such projects has been drawn:

a) Assist local communities to implement an alternative economy to grazing activities such as carpet production and its sustainable distribution trough craftsman cooperatives.

b) Training and assistance in better honey production, or poultry production such as turkeys, etc.

c) Adequate support for the development of the new eco-tourism industry.

d) Micro-financing all activity oriented towards a more sustainable use of the forest habitat including reduction of goats in the forest.

It is clear that this would be only the beginning of the reversal of a trend that needs to be monitored and evaluated step by step, but given the overall picture, there is very little cause for optimism.

References

Camperio Ciani A (1986) La *Macaca sylvanus* in Marocco: sopravvivenza o estinzione. Osservazioni personali e dati storico-demografici *Antropologia Contemporanea* **9** (2) 117-132

Camperio Ciani A (1986) Origine, Evoluzione, Speciazione ed Ecologia del genere Macaca in relazione alle vicende geoclimatiche del Quaternario *Stud. p. l'ecol. del Quater* **8** 9-32.

Camperio Ciani A (1995) Recent decline of the North African Forests: Evidence from biological indicators. In *Biodemography and Human Evolution* pp. 13-16 Ed A Pontecorboli. Florence, Italy

Camperio Ciani A (2002) Demografia e conservazione di *Macaca sylvanus* nel Middle Atlas in Marocco. Workshop: Contributo italiano alla conservazione dei primati: aspetti teorici e pratici. XV convegno Associazione Primatologica Italiana, Roma 30 maggio-1giugno 2002

Camperio Ciani A and Castillo P (2000) The desertification process in the last natural forest of the southern Mediterranean region. In *Proceedings of the International Conference 1996. Volume 2. Summary of project results* pp 471-480 Eds P Balabanis, D Peter, A Ghazi and M Tsogas. Office for Official Publications of the European Communities, Luxembourg.

Camperio Ciani A and Palentini L (2003) La desertificazione in Marocco: uso degli indicatori biologici nel monitoraggio della desertificazione delle foreste del Medio Atlante *Antropologia Mediterranea* **1** n 1 57-68

Camperio Ciani A, Mouna M, e Arhou M (1999) *Macaca sylvanus* as a biological indicator of the cedar forest quality. In *Selected Proceedings of the First International Conference on Biodiversity and Natural Resources Preservation* pp 91-98 Al AKAWAYN Univ. Press.

Camperio Ciani A, Martinoli L, Capiluppi C, Arahou M and Mouna M (2001)

Effects of water availability and habitat quality on bark stripping behaviour in Barbary macaques *Conservation Biology* **15** (1) 259-265

Camperio Ciani A, Palentini L, Arhou M, Martinoli L, Capiluppi C and Mouna M (2005) Population decline of *Macaca sylvanus* in the middle atlas of Morocco *Biological Conservation* **121** 635-641

Caughley G (1977) *Analysis of vertebrate populations.* John Wiley and Sons Ltd.

Deag JM (1977) The status of the Barbary macaque *Macaca sylvanus* in captivity and factors influencing its distribution in the wild. In *Studies In Primate Conservation* pp 267-287 Eds HSH Rainier and GH Bourne. Academic Press, New York, 1977

Drucker GR (1984) The Feeding Ecology of the Barbary Macaque and Cedar Forest Conservation in the Moroccan Moyen Atlas. In T*he Barbary Macaque: A Case Study In Conservation* pp 135-164 Ed JE Fa. Plenum Press, New York

Fa JE (1984) Definition of Age-sex classes for the Barbary macaque. (Appendix 1). In *The Barbary Macaque. A Case Study In Conservation* pp 335-346 Ed JE Fa. Plenum Press, New York

Fa JE, Menard N and Steward PJ (1984) The Distribution and Current Status of the Barbary Macaque in North Africa. In *The Barbary Macaque. A Case Study In Conservation* pp 79-101 Ed JE Fa. Plenum Press, New York

Fellegi IP (1964) Response variance and its estimation *Journal of the American Statistical Association* **59** 1016-1041

Marcharias J (1996) Indagine sui fattori che determinano l'aggressività di *Macaca sylvanus* del Middle Atlas. Ph.D. thesis, University of Florence, Italy

Mehlman P (1989) Comparative Density, Demography, and Ranging Behavior of Barbary Macaques (*Macaca sylvanus*) in Marginal and Prime Conifer Habitats *International Journal of Primatology* **10** (4) 269-292

Menard N and Vallet D (1993) Population Dynamics of *Macaca sylvanus* in Algeria: An 8-Year Study. *American Journal of Primatology* **30** 101-118

Menard N, Quarro M, Latuilliere M, Crouau-Roy B and Le Grelle E (1999) Biodiversity in the cedar-oak forests: the Barbary macaque (*Macaca sylvanus*) as a biological indicator. *Proceedings of the first international conference on biodiversity and natural resources preservation* pp 111-116. Ifrane

Mouna M, e Arhou M and Camperio Ciani A (1999) A propos des population du singe magot (*Macaca sylvanus*) dans le Moyen Atlas. *Proceedings of the first international conference on biodiversity and natural resources preservation* pp 105-109. Ifrane

National Research Council (1981) *Techniques for the Study of Primate Population Ecology.* National Academy Press, Washington

Paul A and Thommen D (1984) Timing of Birth, Female Reproductive Success, and Infant Sex Ratio in Semifree-Ranging Barbary Macaques (*M. sylvanus*) *Folia Primatologica* **42** 2-16

Paul A, Kuester J and Arnemann J (1996) The Sociobiology of Male-Infant Interactions in Barbary Macaques, *Macaca sylvanus Animal Behaviour* **51** (1) 155-170

Sheffrahn W (1999) Habitat fragmentation and genetic variation in *Macaca sylvanus. Proceedings of the first international conference on biodiversity and natural resources preservation* pp 99-103. Ifrane

Scheffrahn W, Menard N, Vallet D and Gaci B (1993) Ecology, demography and population genetics of Barbary macaques in Algeria *Primates* **34** 381-394

Schembri R (2003) Socioecologia delle popolazioni pastorali berbere, valutazione dell'habitat e dell'impatto della pastorizia nelleforeste del medio atlante in Marocco. Dissertation Thesis, University of Padova, Italy

Southwick CH and Siddiqi MF (1988) Partial Recovery and a New Population Estimate of Rhesus Monkey Populations in India *American Journal of Primatology* **16** 187-197

Southwick CH and Siddiqi M F (1994) Population Status of Nonhuman Primates in Asia, with Emphasis on Rhesus Macaques in India *American Journal of Primatology* **34** 51-59

Southwick CH, Richie T, Taylor H, Teas J and Siddiqi MH (1980) Rhesus Monkey Populations in India and Nepal: Patterns of Growth, Decline, and Natural Regulation. In *Biosocial Mechanisms of Population Regulation* pp 151-170 Eds MN Cohen, RS Malpass, HG Klein. Yale University Press, New Haven

Taub DM (1977) Geographic Distribution and Habitat Diversity of the Barbary Macaque *Macaca sylvanus L. Folia Primatologica* **27** 108-133

Taub DM (1984) A Brief Historical Account of the Recent Decline in Geographic Distribution of the Barbary Macaque in North Africa. In *The Barbary Macaque: A Case Study In Conservation* pp 71-79 Ed JE Fa. Plenum Press, New York

Thirgood JV (1984) The Demise of Barbary Macaque Habitat - Past and Present Forest Cover of the Maghreb. In *The Barbary Macaque. A Case Study in Conservation* pp 19 - 69 Ed JE Fa. Plenum Press, New York

van Lavieren E (2004) The illegal trade in the Moroccan Barbary macaque (*Macaca sylvanus*) and the impact on the wild population. Thesis MSc Primate conservation, Oxford Brookes Univ.

Index

www.ingramcontent.com/pod-product-compliance
Lightning Source LLC
Chambersburg PA
CBHW061238220326
41599CB00028B/5470